LIFE HISTORIES OF CENTRAL AMERICAN HIGHLAND BIRDS

PUBLICATIONS OF THE NUTTALL ORNITHOLOGICAL CLUB, NO. 7
Editor, Raymond A. Paynter, Jr.

LIFE HISTORIES OF CENTRAL AMERICAN HIGHLAND BIRDS

by
ALEXANDER F. SKUTCH

CAMBRIDGE, MASSACHUSETTS
Published by the Club
1967

PUBLICATIONS OF THE NUTTALL ORNITHOLOGICAL CLUB

1. The Comparative Biology of the Meadowlarks (*Sturnella*) in Wisconsin. Wesley E. Lanyon, 1957. 67 pp., 31 pls. Cloth $2.00.
2. Comparative Breeding Behavior of Four Species of North American Herons. Andrew J. Meyerriecks, 1960. 158 pp., 15 pls. Cloth $3.00.
3. Structural Adaptations of the Head and Neck in the Black Skimmer, *Rynchops nigra* Linnaeus. Richard L. Zusi, 1962. 101 pp., 44 figs. Cloth $3.00.
4. The Rough-winged Swallow *Stelgidopteryx ruficollis* (Vieillot). A Study Based on Its Breeding Biology in Michigan. William A. Lunk, 1962. 155 pp., 19 figs., 3 pls. Cloth $4.00.
5. Ecology and Bioenergetics of the Long-billed Marsh Wren, *Telmatodytes palustris griseus* (Brewster), in Georgia Salt Marshes. Herbert W. Kale II, 1965. 142 pp., 61 tables, 22 figs. Cloth $4.00.
6. Communication and Relationships in the Genus *Tyrannus*. W. John Smith, 1966. 250 pp., 3 tables, 51 figs. Cloth $6.00.
7. Life Histories of Central American Highland Birds. Alexander F. Skutch, 1967. 213 pp., 1 table, 6 figs. Cloth $6.00.

Printed in the United States of America

Publications of the Nuttall Ornithological Club may be obtained from the Nuttall Ornithological Club, c/o Museum of Comparative Zoology, Harvard University, Cambridge, Massachusetts 02138.

CONTENTS

CONTENTS

INTRODUCTION

The "highland birds" included in this book are chiefly species whose center of abundance lies at altitudes of more than 3,500 feet above sea level. Many of them live only at far greater heights; a few range upward from the lowlands but are as abundant around 4,000 feet as I have found them anywhere. Each of the 39 species is treated as fully as the present state of our knowledge of its habits permits. For some I am able to give fairly complete biographies; others, less thoroughly studied, are included for their arresting songs, their unusual nidification, or some other noteworthy aspect of their lives.

Inadequately as the habits of the birds of the tropical American lowlands are known, those of the highlands are far less known. The naturalists who have been interested in the life histories of tropical birds, rather than in collecting their skins, have spent much more time in the warm lowlands than in the cool, elevated inland regions. Most of the biological stations in tropical America where ornithologists have worked have been at low altitudes. The New York Zoological Society's station at Kartabo, British Guiana, where William Beebe and his associates made extensive studies early in the present century, was only 25 feet above sea level. Only slightly higher is Barro Colorado Island in Gatún Lake in the Panama Canal Zone, whence over the years we have had a succession of important ornithological studies by Frank M. Chapman, Josselyn Van Tyne, Alfred O. Gross, Martin H. Moynihan, and others. The New York Zoological Society's present station at Simla in Trinidad, where David W. Snow has recently made outstanding studies, embraces altitudes from 500 to 1,800 feet above sea level. Considerably higher is the Estación Biológica de Rancho Grande in the state of Aragua, Venezuela, where the main building, at 3,575 feet, is situated in a nature reservation which extends upward to the summit of the coastal range at 7,875 feet. Here Ernst Schäfer made some splendid studies of highland as well as predominantly lowland birds.

Of ornithologists who have made notable life history studies at places in tropical America other than biological field stations, David E. Davis in Cuba, Edwin Willis in British Honduras, F. Haverschmidt in Surinam, Helmut Sick and Cory T. de Carvalho in Brazil, Stephen Marchant in Ecuador, and others have worked largely or wholly in the lowlands. Jean Dorst, however, chose for the scene of his investigations the high puna of southern Peru, where at altitudes above 13,000 feet the environment to which living things must adapt differs more from that at moderate altitudes than the latter from the warm lowlands. At similar heights in the Andes of Peru, Anita K. and Oliver P. Pearson made a careful study of tinamous. Away to the north, in Mexico, Helmuth O. Wagner has made important observations on a variety of highland birds, especially hummingbirds.

Of the 32 nesting seasons that I have passed with tropical birds, most were spent at altitudes between 2,000 and 3,000 feet, where one finds the majority of the non-aquatic lowland birds of a region with a blessed mitigation of the lowland heat and its attendant diseases. I have passed four seasons near sea level, and four in the highlands as I have somewhat arbitrarily defined them for the purposes of the present book. In addition to the more detailed studies made at the four highland localities where I resided for a breeding season or longer, much information about highland birds, especially their altitudinal distribution and behavior when not nesting, was gathered in other places that I visited more briefly, usually in the off season. The nine localities where my field work in the highlands was chiefly done, the altitudes, the dates, and the character of each locality, were as follows:

1. Sierra de Tecpam (also spelled Tecpán), Department of Chimaltenango, west-central Guatemala, 8,000-10,000 feet, 12 November to 2 December 1930, and 5 January 1933 to 3 January 1934.

From the 7,000-foot plateau on which the town of Tecpam is situated, this range rises to 10,000 feet at its highest point. Most of my time on the Sierra was spent on the estate known as "Chichavac," about midway between its base and its crest. Here, at an altitude around 8,500 feet, was a topography that sometimes reminded me of the Piedmont region of Maryland where I passed my boyhood. Level or gently inclined fields alternated with low ridges wooded by oaks and pines. Only when I reached the edge of one of the deep, trench-like gorges or *barrancos* that here and there dissected the terrain, or came to an elevation whence I could look over the plateau, far below, to the great conical peaks that dominated the horizon, was I forcibly reminded that I was in volcanic Guatemala. The vegetation, too, here in the altitudinal temperate zone, helped to sustain the illusion that I was somewhere beyond the tropics. Beneath the pines and oaks, and on the open meadows, grew violets, buttercups, lupines, thistles, chickweeds, clovers, selfheal, plantains *(Plantago)*, and many weeds hard to distinguish from those in northern lands. Many of the shrubs and trees also represented familiar northern genera. There were no palms on these cool heights; but epiphytes grew on the older trees, especially in sheltered valleys, with a profusion that one finds only in the tropics, and the presence of arborescent fuchsias, dahlias, begonias, and many less familiar plants of tropical groups, which I sometimes collected with numb fingers, helped me to keep in mind that, although uncomfortably cold, I was in a tropical country.

The bird life presented the same stimulating mixture of northern and tropical types. Along with siskins, Spotted Towhees,[1] Common

[1] The scientific names of all species of birds mentioned in the text are listed in the index after the English name.

Bluebirds, Steller's Jays, Greater Pewees, Red-shafted Flickers, Hairy Woodpeckers, and Whip-poor-wills were such distinctly tropical forms as trogons, motmots, toucans, antbirds, cotingas, shrike-vireos, and honeycreepers. The Brown Creepers and the Spotted-crowned Woodcreepers, which climbed up the same trunks, so closely resembled each other in form, coloration, and mode of foraging that it was hard to believe that they were not closely related, that one represented the Holarctic, the other the Neotropical, avifauna.

Above 9,000 feet on the Sierra de Tecpam the oaks gave way to cypress *(Cupressus Benthamii)*, of which in 1933 there were large, scarcely touched tracts of magnificent old trees. Here my hosts had another estate known as "Santa Elena," where I spent all of January, with shorter visits at other seasons. Here were Golden-crowned Kinglets, Yellowed-eyed Juncos, and Horned Guans, all of which were rare or absent lower on the range. Yet Blue-throated Green Motmots, Spotted-crowned Woodcreepers, Banded-backed Wrens, and Cinnamon-bellied Flower-piercers endured the nocturnal chill of these heights, and of course there were many hummingbirds.

The rainy season of 1933 broke in mid-May and turned out to be exceptionally severe, causing disastrous landslides that buried or carried away highways and wrecked the one railroad that penetrated the western highlands of Guatemala. The dry season began in mid-October and lasted seven months. Although through most of this long interval there were occasional stormy days, when the Sierra was enveloped in wind-driven clouds and light rain fell, the earth became increasingly dry; by March much of the herbaceous vegetation had withered, except in deep woodland and moist ravines, and there were few flowers. The nocturnal frosts, which began as soon as the returning dry season brought clear night skies and continued until the beginning of April, killed the shoots of the more tender herbs that were exposed to it, although they had little effect on the higher woody plants. For five months, dawn at the end of every clear and windless night revealed the pastures and open fields white with frost, which crunched underfoot as I walked over it to begin the day watching the birds. As the rising sun warmed the air, I would gradually shed the thick clothes that I wore at daybreak; in March the midday air was so hot and dry that my lips tended to crack. The desiccated ground cover on even moderate slopes became so slippery that I could hardly climb them without rubber soles.

Most of the birds on the Sierra de Tecpam nested in the six weeks of mild, dry weather between the cessation of nocturnal frost at the beginning of April and the advent of the hard, cold rains in mid-May. The dominant oaks, along with the abundant tall alders and many other trees, performed a seeming miracle by renewing their foliage at the height of the dry season, and their new leaves supported enough insect life to nourish the breeding birds and their young. Those birds that foraged on the ground waited until somewhat later, and few had nestlings to feed before the returning rains had soaked the forest litter and the soil beneath it, reactivating the small creatures on which these birds fed. The young of some of these late breeders died in the nest, apparently because of unfavorable weather. Nectar-feeders, including the hummingbirds and an aberrant honeycreeper, the Cinnamon-bellied Flower-piercer, had a breeding season all their own, at the beginning of the dry season, when the sunny skies and the earth still moist from months of rain brought forth a profusion of blossoms that brightened all the highlands. Most kinds of birds reared only a single brood, no larger than that of closely related species of the lowlands, where the breeding season may be much longer. More details on the breeding seasons of highland birds are given elsewhere (Skutch, 1950a).

In my 13 months on the Sierra de Tecpam, I indentified 123 species of birds between 7,000 and 10,000 feet. Of these, 34 were long-distance migrants from the north, leaving 89 species that breed in Guatemala. Among the local birds, there was a certain amount of altitudinal migration. Green Violet-ear Hummingbirds and Black Thrushes, very evident in their respective nesting seasons, vanished from the Sierra after breeding was over. Red-eyed Cowbirds were noticed on the Sierra only from March until July, when other kinds of birds provided an abundance of nests into which they could drop their eggs; at other times, these parasites preferred the open plains at the foot of the range. White-winged Doves, which favor somewhat arid country, nested on the Sierra toward the end of the dry season but vanished after the rains became heavy, to return with drier weather in the following November. Other species seemed to wander upward after nesting at lower altitudes. Common Bush-Tanagers and Blue-Hooded Euphonias, rarely seen during the first half of the year, became abundant near the crest of the range in late July. Scaled Antpittas were found among the lighter oak woods in the wet season, but after the advent of drier weather they seemed to have migrated upward to the more humid cypress forests of the mountaintop. Some species kept very strictly to a narrow altitudinal belt: Golden-crowned Kinglets, not rare among the cypresses above 9,000 feet, were never seen below this.

2. From July to December, 1934, I travelled extensively in the western highlands of Guatemala, in the departments of Quezaltenango, Huehuetenango, and El Quiché, while collecting plants for the Arnold Arboretum of Harvard University. These journeys, largely on horseback, while allowing no time for detailed studies of birds, permitted numerous observations on their habits and altitudinal distribution. In contrast to the wet season of 1933, that of 1934 was, fortunately, exceptionally mild.

3. San Miguel de Desamparados, at the foot of the Tablazo Moun-

tains south of the Meseta Central of Costa Rica, 9 September to 18 November 1935. The little country hotel, at about 4,500 feet above sea level, where I lodged while engaged in writing, was surrounded by coffee plantations, pastures, and cultivated fields with a depauperate avifauna; but by climbing up to the summits of the Tablazo, at about 6,500 feet, I reached patches of mossy subtropical forest where I became acquainted with several species new to me.

4. Montaña Azul, two miles below Vara Blanca de Sarapiquí, Province of Heredia, Costa Rica, 5,000-6,000 feet, 8 July 1937 to 11 August 1938, with absences from 11 November to 7 December and 14 December to 12 January.

The cottage, which, with the help of friends, I persuaded a German coffee planter on the Meseta Central to rent to me, was situated, at an altitude of 5,500 feet, on the back of a narrow ridge or spur which on three sides fell off abruptly into the ravines characteristic of the deeply dissected northern slopes of Costa Rica's volcanic Cordillera Central. To the east, this spur dropped precipitously hundreds of feet into the profound gorge through which flowed the Río Sarapiquí. The top of the ridge had been cleared and sowed with pasture grasses, but its steep sides were still wooded; and to the north and east the forest, with its natural fauna hardly disturbed by man, stretched on and on to the limits of vision. A new clearing so situated is a precious boon to the birdwatcher, for in the surrounding treetops he can follow the movements of birds that he could hardly glimpse from the depths of unbroken forest, and the dead or dying trees in its midst are irresistibly attractive to hole-nesting birds, which emerge from the adjoining woodland to rear their families in full view.

In these subtropical forests, the trees grew impressively high wherever the ground was level and stable enough to support them. Especially noteworthy were the massive cedars *(Cedrela)* that grew across the gorge where they could not easily be lumbered. Wild avocados *(Persea Schiedeana)* also attained an impressive size. The confusing variety of lauraceous trees provided fruits that were attractive to Quetzals, Three-wattled Bellbirds, Lovely Cotingas, Bare-necked Umbrellabirds, Blue-throated Toucanets, and other species. Oaks were absent from Montaña Azul but grew to magnificent proportions a few hundred feet higher up the slopes. Nowhere have I seen a greater profusion of epiphytes than on these storm-beaten heights. Some of the larger trees appeared to be burdened with tons of these aerial growths, which ranged in size from smaller trees down to mosses, liverworts, and doubtless a microscopic flora that would well repay study. The whole gamut of tropical epiphytes was represented in these aerial gardens: ferns, orchids, aroids, bromeliads, clusias, and a fine variety of heaths. In the pastures, the larger stumps, which had been cut high to avoid the spreading buttresses at their bases, were so capped with epiphytic growths that they resembled gigantic green toadstools. In contrast to the situation in the Guatemalan highlands, few of the woody plants represented genera that I had known since I was a boy. To learn their classification, it was necessary to make specimens, send them to distant museums, then wait for months or years until I received their names.

Among the resident birds, too, were far fewer of northern affinities. Even the widespread Hairy Woodpecker had here become so dark that a visitor from the north might take it for a different species. In the more open country near the continental divide, Common Meadowlarks uttered their familiar song, but they had not yet colonized this more recent clearing in the forest. On the other hand, such tropical families as trogons, toucans, woodcreepers, cotingas, and tanagers were well represented; and here for the first time I made the acquaintance of a number of Andean genera, especially in the ovenbird family. These strange brown inhabitants of damp subtropical forests were challenging but infinitely elusive, as were the only species of Rhinocryptidae north of Panama, the Silver-throated Tapaculo, and the little Wren-Thrush, long considered to be the unique representative of the family Zeledonidae. In the course of a year, I identified 156 species of birds between 5,000 and 6,000 feet. Of these, 24 were long-distance migrants from the north, leaving 132 that breed in Costa Rica. Thus the avifauna was considerably richer than that on the Sierra de Tecpam. Here, as elsewhere, I failed to identify swifts (except the easily recognized White-collared Swift) that I saw only while they circled high overhead; nor several hawks that I glimpsed only once or twice.

At this altitude, I was below the frostline in Costa Rica, yet much of the time the weather was far from balmy. Throughout the year, but especially during the months of the northern autumn and winter, there were storms when for days together the northeast winds, sweeping from the Caribbean Sea across the lowlands of eastern Nicaragua and northern Costa Rica, blew the rain clouds unrelentingly against this north-facing slope, blotting out the sun, enveloping all the trees in gray mist, and causing everything they touched to drip with moisture. I considered myself fortunate, in this then remote region, to have obtained a house with double walls. But the unpainted boards, nailed up while still unseasoned, had shrunk, leaving wide gaps, through which, in the worst of the storms, the wind drove the rain, making the northern side of the cottage scarcely habitable. The rosy-cheeked mountaineer children stood this grim weather surprisingly well; but I had become acclimated to lower altitudes in the tropics, and after several sunless days I found the raw dampness terribly depressing. The birds that I most wished to study, especially the Quetzals, the Three-wattled Bellbirds, and the Blue-throated Toucanets, became so rare in the stormiest period of the year that it was difficult to preserve faith that I would find their nests in due season. Yet when the storm finally blew itself out and the sun shone forth in the bluest of skies, illuminating all the bright colors of the flowering epiphytes and glinting from the glossy leaves in a myriad spears of light, I could easily imagine that I had succumbed to the storm and been resurrected in a naturalist's heaven.

As in the Guatemalan highlands, the birds at Montaña Azul nested chiefly in April and May, except the nectar-drinking hummingbirds and flower-piercers, which as usual were a law unto themselves. But here, instead of breeding in a dry period, they did so in very wet weather, when sometimes the cold rain continued so long that I feared the nestlings would perish; but the hardy parents, accustomed to such storms, brought them through alive in the nests that I watched. A few species, including the Quetzal and, surprisingly, the Blue-throated Toucanet, raised two broods, which prolonged their breeding into July and even August. But here, as in other highland localities of Central America where I have worked, there was a sharp falling off in the number of new nests after the end of May. In 65.8 per cent of the 152 nests that I found at Montaña Azul, eggs were laid in April and May, an equal number in each of these months. On the Sierra de Tecpam, where I discovered only 88 nests, eggs were laid in 31.8 per cent in April and in 21.6 per cent in May, giving 53.4 per cent for the two months when nesting was at its height. The difference between the percentages for the two localities is due largely to the fact that I found more nests of nectar-drinkers, especially hummingbirds, on the Sierra de Tecpam than at Montaña Azul. On the Sierra, the nests of these flower-visitors found from October to January accounted for 26.2 per cent of the whole number of nests. With the exclusion of these, the breeding season at the higher altitude becomes still more sharply peaked in April and May (details in Skutch, 1950a).

Although there were not so many long-distance migrants from the north at Montaña Azul as on the Sierra de Tecpam, among its richer bird life more altitudinal movement was noticed. A number of species were first recorded as the breeding season approached, eight or nine months after my arrival. Others vanished during a long-continued rainstorm in mid-July and were not seen again until December or later. Among these wanderers were:

Three-wattled Bellbird (*Procnias tricarunculata*)—vanished in July; reappeared in December; prominent in the breeding season.

Chestnut-headed Oropéndola (*Zarhynchus wagleri*)—first seen on 19 February; small colony started building in April.

Masked Tityra (*Tityra semifasciata*)—left in August; not seen again until 20 February; nested.

Blue Tanager (*Thraupis episcopus*)—left in August or September; reappeared 21 February; probably nested.

Gray-capped Flycatcher (*Myiozetetes granadensis*)—left in August. A pair appeared on 25 February and nested.

Vermilion-crowned Flycatcher (*Myiozetetes similis*)—first seen at Montaña Azul on 25 February; probably nested.

Mountain Elaenia (*Elaenia frantzii*)—first noticed at Montaña Azul on 3 March; nested.

Black-striped Sparrow (*Arremonops conirostris*)—disappeared in early August; next seen on 3 March; probably nested.

Swallow-tailed Kite (*Elanoides forficatus*)—arrived 5 March and nested.

Sulphur-bellied Flycatcher (*Myiodynastes luteiventris*)—left in mid-July; reappeared 7 March; nested.

Squirrel Cuckoo (*Piaya cayana*)—first seen in mid-March; probably nested.

Red-eyed Cowbird (*Tangavius aeneus*)—first seen 28 March; disappeared in wet spells, to reappear in drier intervals.

Tropical Kingbird (*Tyrannus melancholicus*)—appeared in early April.

Yellow-bellied Siskin (*Spinus xanthogaster*)—first seen 5 April; nested.

Piratic Flycatcher (*Legatus leucophaius*)—first seen 6 April; probably nested.

Giant Cowbird (*Psomocolax oryzivorus*)—first seen 6 May at the colony of oropéndolas, which these cowbirds parasitize.

Of the birds which seemed to be absent from Montaña Azul in the fall and part or all of the winter but returned to pass the breeding season there, the Swallow-tailed Kite, the Sulphur-bellied Flycatcher, and the Piratic Flycatcher are not found anywhere in Costa Rica in the winter months and evidently pass this season in South America. Other birds apparently moved either up or down the mountainside. The Blue Tanager, the Vermilion-crowned Flycatcher, and the Tropical Kingbird probably came from the more open country higher up the mountain slopes (nearer the deforested Central Plateau), where I saw them while they were absent from Montaña Azul. Most of the other wanderers had apparently come up from lower and warmer levels to the north. The Three-wattled Bellbird is often present at lower altitudes through much of the year and even sometimes in the breeding season, but it seems not to nest much below 5,000 feet.

Montaña Azul proved to be such a richly rewarding locality that I have often regretted that I passed only a single nesting season there. But the botanical collecting on which I then depended for an income made it expedient to shift to a fresh locality at the end of a year. When, after an interval of 25 years, I revisited Montaña Azul, coming in an automobile over a busy paved highway instead of trudging for six miles along one of the miriest roads I have ever known, the beautiful forests had been converted into pastures; Quetzals and Black Guans had been replaced by Rufous-collared Sparrows and other widespread birds of open country. The pretty little cottage that I had occupied had been converted into a laborers' barracks, blackened with smoke.

5. Hacienda "Las Cóncavas," between Cartago and Paraiso, Costa Rica, 4,000-4,500 feet; frequent visits, mostly short, between 1935 and 1954, the longest 13 August to 16 September 1938.

On this hospitable farm the land was occupied by coffee, pastures, cultivated fields, and copses of native or planted trees. The avifauna consisted chiefly of the common, widespread birds of open country at middle altitudes, with a few challenging but elusive thicket-haunting species.

6. Baños and vicinity, Province of Tungurahua, Ecuador, 4,000-8,500 feet, 26 September to 29 October 1939.

The good localities for birds were down in the Pastaza Valley or far up on the slopes of Volcán Tungurahua, too distant from my lodging in the town for life history studies. At this season, a number of birds were nesting, and others were feeding fledglings.

7. Hacienda "La Giralda," a dairy farm at the western end of the Barba massif in the Cordillera Central, near the hamlet of Los Cartagos, Province of Heredia, Costa Rica, 6,500-8,000 feet, 25 February to 6 July 1963.

The house that we occupied at La Giralda looked over the western part of the Central Plateau toward the Pacific. On exceptionally clear days the Gulf of Nicoya, with its islands and the mountainous peninsula beyond it, were visible in the far distance, while at night the lights of Alajuela and a number of smaller towns of the plateau twinkled brightly far below. The long slopes over which I roamed looking for birds were occupied chiefly by pastures, which for careful management had been divided by fences into small plots, so that one could not walk far without opening and closing many gates or squeezing through the barbed wire every hundred yards or so. These pastures were shaded by numerous trees, which in the lower reaches were chiefly native alders (*Alnus acuminata*) and cypresses (*Cupressus Benthamii*) introduced from Guatemala or Mexico. In the higher pastures, the more varied shade trees were of native species, especially Winter's bark (*Drimys Winteri*), a cornel (*Cornus disciflora*), and several kinds of myrtles. In the many ravines that intersected the grassy slopes were patches of primary forest or second-growth woods, from less than an acre to several acres in extent, which had been left to protect the watershed. To the northwest of the pastures was a long, deep valley whose sides were covered with several hundred acres of forest that was dominated by huge, epiphyte-laden trees, including many oaks. The steepness of these wooded slopes of the farm, and the dense undergrowth of tall, cane-like bamboos, made it difficult to move around on them and watch the birds.

Many of the birds at La Giralda were the same as I had found at

Montaña Azul, just across the continental divide and about 6 miles away in an airline. Others, which barely reached Montaña Azul, were more or less common at this higher altitude, including the Long-tailed Silky-Flycatcher, Flame-throated Warbler, Black-cheeked Warbler, and Flame-colored Tanager, of each of which I found one or more nests. And of course many species that bred at Montaña Azul did not reach this height. The Quetzal, abundant there in the nesting season, was here rare and elusive. Partly because of the greater altitude, and partly because of the reduced area of forest and greater disturbance by man, the avifauna here was far less varied than I had found it at Montaña Azul a quarter of a century earlier. In four and a quarter months I identified 86 species, of which 15 were long-distance migrants from the north, leaving only 71 species that breed in Costa Rica. As at all wooded localities in Central America, the warbler family provided the greatest number of migratory species.

I saw no frost at La Giralda, but was told that earlier in the year it whitened the fields. These mountain slopes drained into the Pacific, but they were such a short way below the continental divide that the Caribbean weather "spilled over" to them. The strong, persistent northeast winds that had so often brought wet and gloomy days to Montaña Azul made themselves felt here. Indeed, as they were now blowing downward, they seemed even stronger; but much of their moisture had been dropped while crossing the crest of the range and they carried less rain. Until the end of March, these northerly winds prevailed, often blowing steadily all day, and sometimes attaining such force that they broke branches from the trees, making birdwatching not only unprofitable but even somewhat perilous. These winds often drove the clouds and drizzle through the trees, but in March they brought little rain; and the dry season continued until early April, when the winds from the Caribbean gave way to those from the Pacific, which blew more gently but brought thunderstorms and heavy downpours. Yet even after the rainy season was well established, there were days when the northeast wind returned and blew such a gale that I feared it would tear from the trees the nests of the Long-tailed Silky-Flycatchers, the birds that I chiefly studied at La Giralda.

Whenever it veered more to the east, the wind brought fine volcanic cinders from Irazú, the huge sprawling volcano 20 miles away, which began to erupt in late March of 1963 and continued until toward the end of the following year. Often there was a thin deposit of this ash on the foliage and pasture grass, but here it never became deep nor caused much damage, as happened on the Central Plateau southwest of the crater.

Although I found a few birds nesting in March, breeding did not become general until the following month. Most of the birds raised their families in April and May, when on many days there were only a few hours of watery sunshine in the morning, before the clouds drifted up the mountainside to obscure the sun, and hard showers fell in the afternoon; or else the chill gray cloud-mist shrouded the mountain slopes from dawn to nightfall. Yet occasionally there was a fine rainless day in these months. The majority of the birds were evidently single brooded, and by early July few were still engaged with nests. On the whole, the volcanic ash that from time to time dusted the foliage seemed to have little effect on their breeding.

8. Cañas Gordas-San Vito de Java region, Pacific slope of Costa Rica near the Panamanian border, 3,500-4,000 feet, 14 March to 24 June 1964.

Most of this interval was spent at Finca "Loma Linda," between Cañas Gordas and Agua Buena. The pastures and coffee plantations which occupied the higher parts of this beautiful farm had been made without the usual felling and burning, so that many noble trees of the original forest remained standing in them, especially in the coffee plantations, where they provided the shade that this shrub requires except at the highest altitudes at which it thrives. A majestic Mexican elm (*Chaetoptelea mexicana*) with widely spreading buttresses, standing in the central pasture, was measured by triangulation and found to be 170 feet high. The more abundant maría (*Calophyllum brasiliense*) was lower and more slender, but still an impressive tree. In the tall, dense forest on the slopes below the clearings grew the campana (*Laplacea semiserrata*), cerilla (*Symphonia globulifera*), a variety of lauraceous trees, and innumerable palms (*Euterpe*) with feathery crowns upheld on tall, gray, elegantly slender, columnar trunks. Many of the trees bore a luxuriant growth of epiphytes, and moss grew thickly on limbs and trunks. The undergrowth on these wooded slopes was dense, with the usual profusion of shrubs and small trees belonging to the Piperaceae, Melastomaceae, Rubiaceae, and other families, and an abundance of tree ferns, lower ferns, and large-leafed herbs related to the banana and the ginger plant. The deep, narrow ravines into which these slopes fell were so cluttered with fallen trunks and branches that to move along them was toilsome. Had it been feasible to explore these ravines extensively, they would doubtless have yielded many rare nests.

The richness of the avifauna is indicated by the fact that within a radius of a few miles, and an altitudial range of only 500 feet, I identified in three and a half months 206 species—and this in an area which provided no habitat for waterbirds, except the slender Green Kingfisher, which is satisfied with a narrow forest stream. Of these 206 species, 27 were long-distance migrants from the north, leaving 179 species that nest in Costa Rica. Nearly three-quarters of these local birds ranged upward from sea level, indicating clearly that this region is in the Tropical Zone. A number of species that I found here avoid the coastal lowlands and are most abundant in the higher parts of the Tropical Zone. In this class are the Green Hermit Hummingbird, Redheaded Barbet, Spotted Woodcreeper, Plain Antvireo, Scaly-crested Pygmy Flycatcher, Orange-billed Nightingale-Thrush, Scarlet-thighed Dacnis, Golden-crowned Warbler, and Gray-striped Brush-Finch. Among the subtropical birds that I found between 3,500 and 4,000 feet in the Cañas Gordas district were the Spotted Wood-Quail, Blue-throated Toucanet, Ochraceous Wren, and Black-faced Solitaire. It was interesting to find the altitudinal ranges of the Highland and the Lowland Wood-Wrens overlapping here. The lowland species lurked in the undergrowth of the less humid forests on the ridges; the Highland Wood-Wren sang in the cool, damp ravines below. Probably somewhat more subtropical species would be found at this altitude where the mountain slopes sweep directly up to subtropical heights; but the moderate elevations of the Cañas Gordas district are separated by slightly lower land from the high Cordillera de Talamanca far in the north.

Here, as in more elevated parts of Central America, the nesting season was at its height at the beginning of the rainy season, in April and May; but probably the peak of breeding activity in these months is less sharp than in the subtropical and temperate zones. Although I found a few nests in March, the main flush of breeding came later in April than I expected, probably because in 1964 the unusually prolonged dry season continued well into this month. Nevertheless, by late June, when we left, active nests had become hard to find.

9. Río Cotón, Zeledón Estate, Pacific slope of Costa Rica near the Panamanian border, 4,700-5,000 feet, 15 to 23 February 1965.

Although we spent only eight days here, the visit was important to me because I found a number of birds at a greater altitude than I had ever seen them before. The forest which for miles surrounded the small clearing in which we camped, beside the clear, cool Río Cotón, presented a number of peculiar features. It was dominated by an extraordinary abundance of huge, spreading fig trees, whose great irregular trunks, composed of coalescing aerial roots, often had a central hollow, left by the decay of the host tree they had enveloped and strangled. Considering the altitude, there was a surprising paucity of palms large and small. The tall trees bore far fewer epiphytes of all kinds than one expects at this height in the Costa Rican mountains. Finally, although the shrub layer of the forest was well developed, the herbaceous ground cover was remarkably sparse. This woodland was easy to walk through, except where the ground was occupied by thickets of tall, cane-like bamboos, which were everywhere flowering and setting seed, after which they would die. The forester of our party, Leslie Holdridge, attributed these peculiarities of the forest to the pronounced dry season; he called it a "monsoon forest."

Here, in a week, Paul Slud and I identified 114 species of birds, of which 17 were migrants from the north. Doubtless a number of the resident birds escaped us, for part of the time a persistent strong east wind made bird finding difficult; and few had yet begun the nuptial singing or calling which often reveals their presence in heavy forest where when silent they lurk unseen. Despite the presence here of certain Tropical Zone birds higher than I had found them elsewhere, less than half of the resident species ranged up from sea level, and these were chiefly the hardier, more adaptable kinds, or else species whose center of abundance is not in the lowlands but in the upper part of the Tropical Zone. Here again both the Highland and the Lowland Wood-Wrens were present, indicating a wide overlap of these congeneric birds characteristic of different life zones. But the Streaked-headed Woodcreeper, which was not rare at Cañas Gordas, here seemed to be wholly replaced by its highland counterpart, the Spotted-crowned Woodcreeper. The subtropical element of the avifauna was strong on these southern slopes of the Cordillera de Talamanca at 5,000 feet.

Altitudinal Distribution

In the following accounts, I give the altitudinal distribution of the birds, so far as known to me from published records or personal observations, in feet rather than in meters, because I believe that most English–speaking people remember the heights of mountains, cities, and other geographical points in the units with which they were familiar as children rather than those of the metric system. I give the altitudinal distribution in terms of height above sea level rather than life zones for several reasons. In the first place, this will avoid confusion, because different writers have used the names of altitudinal life zones in different senses: the tropical and subtropical zones of Slud (1964), for example, are not those of Todd and Carriker (1922) or of Griscom (1932). Secondly, knowing a bird's altitudinal distribution in terms of feet or meters above sea level tells us more clearly where we can expect to find it than knowing the life zone(s) in which it occurs; because it is easy to determine one's altitude, within a few hundred feet, with a pocket aneroid altimeter or a good topographic map, whereas to know in what life zone one is situated is frequently perplexing. Finally, if, as I believe to be the proper course, we are to establish our life zones upon biological rather than meteorological data, we must first learn the altitudinal distribution of the greatest possible number of organisms.

There is no doubt that the concept of life zones, latitudinal and altitudinal, crystallizes an important truth about the distribution of vegetable and animal life on this planet. But as usually happens, when we try to fit the infinite diversity of nature into our neat conceptual pigeonholes, we run into serious difficulties. The novice who sees life zones drawn on a map might imagine that their limits are as sharply defined as the artificial boundaries between states. Nothing could be further from reality. Only where the topography is very abrupt, as on Mt. Duida, Venezuela, with its long line of towering cliffs, are life zones sharply delimited. Where mountain slopes sweep upward for thousands of feet with no major break in their continuity, the vertical life zones fade into each other; they interdigitate like cogwheels. The problem of delimiting them by biological rather than meteorological data is aggravated by the apparently capricious way in which birds and other organisms distribute themselves altitudinally. Of two species of land birds which mingle at sea level, one will drop out in the first thousand feet or so of ascent, while the other continues upward to 8,000 or 9,000 feet. Similarly, of two species which live together on a mountaintop at 10,000 feet, one will be left above us as we descend below 9,000 feet, while the other accompanies us downward to 5,000 feet or even lower.

If I had adequate data and attempted to delimit the life zones of a mountainous tropical country, I should first of all plot the distribution of a large number of species on a vertical scale, representing each by a spindle-shaped figure fattest at the bird's center of abundance and tapering off at the ends. Then I should draw horizontal lines, the boundaries between zones, at levels where they would intersect the smallest number of these vertical figures, or at least avoid intersecting them where they are fattest. But I am sure that it would be impossible to place these zonal boundaries at any level where they would not cut squarely across the center of abundance of certain species. When we are in the midst of a life zone, we can usually recognize it without difficulty; but between every pair of adjacent zones there is typically a broad band of transition in which it is perplexing to tell just where we are. What we are dealing with is a continuous spectrum of change, in which certain regions can conveniently be named, rather than discrete, superimposed strata, such as we sometimes find in a geological formation. Life zones give us useful descriptive terms, but only exceptionally do they provide the neatest way of stating the distribution of some particular organism.

Since Chapman (1917) published his study of the distribution of bird life in Colombia, it has been customary among ornithologists to recognize four altitudinal life zones in tropical America: the Tropical, the Subtropical, the Temperate, and, from Costa Rica southward, the Páramo. Each altitudinal zone can be conveniently divided into a humid and an arid section, reflecting genuine differences in the composition of the avifauna.

In his pioneer work on the birds of Costa Rica, Carriker (1910) tried to establish life zones in accordance with the probable origin of the avifauna (whether northern or tropical) as well as its present distribution; and this double criterion resulted in a confusing multiplicity of overlapping zones. Later he and his coworker (Todd and Carriker, 1922), dealing with the avifauna of the Santa Marta region of Colombia, abandoned this too-involved system in favor of Chapman's simpler concept that deals more effectively with the realities of vertical distribution in the tropics. Recently Slud (1964) has considered the distribution of Costa Rican birds in accordance with a more complicated system of zones, established by meteorological data, and much used by plant ecologists and foresters; but perhaps these zones are too finely divided for tropical ornithology, at least in its present state of development. Although nobody doubts that climate affects bird life, both directly and through its influence on vegetable and insect life, the question which ornithologists must decide—and nobody else can decide it for them—is just where in a climatic gradient the effect upon the composition of the avifauna becomes so pronounced that it is desirable to recognize a different zone.

For Guatemala, Griscom (1932:33) gives the vertical extent of the life zones as follows:

Tropical Zone—sea level to 3,000-4,500 feet;
Subtropical Zone—3,000 to 6,000 feet and locally much higher;
Temperate Zone—5,000 to 13,000 feet.

For the Santa Marta region of northern Colombia, Todd and Carriker (1922:58) assign these extensions to the life zones:

Tropical Zone—0 to 4,500 feet;
Subtropical Zone—4,500 to 9,000 feet;
Temperate Zone—9,000 to 11,000 feet;
Páramo Zone—11,000 feet to snow-line (15,000 feet).

So far as it is possible to assign precise limits to something so broad and ill-defined as the boundaries (or, more properly, belts of transition) between life zones, the scheme adopted for the Santa Marta region, which is slightly farther to the south, applies fairly well to Costa Rica. Of course, the Costa Rican mountains do not reach snowline (the highest, Cerro Chirripó, being only 12,580 feet), and their páramos are too small to have developed a distinctive avifauna and are too isolated to have acquired one from the extensive Andean páramos.

The highland birds treated in this book live in the Temperate Zone, the Subtropical Zone, and the upper quarter of the Tropical Zone; a few descend to the lower limit of this zone, at sea level. A fact that we shall repeatedly notice in the following accounts is that birds of the higher life zones descend to lower altitudes on the exposed, windward, Caribbean slopes of the Cordillera Central of Costa Rica than they do on the more protected, southern, Pacific slopes of the Cordillera de Talamanca. This fact may be epitomized by saying that the life zones, above the Tropical, are tilted downward from the drier, leeward side of the country to the wetter, windward side. The reason for this is, no doubt, that wet, cloud-bathed slopes, and the deep ravines they embrace, offer, at lower altitudes than elsewhere, the cool, humid conditions that many Subtropical birds prefer and many Tropical Zone birds avoid.

Another fact worthy of notice is that, as one ascends mountains in tropical America, from Costa Rica southward, the avifauna becomes poorer in species. This is shown by a comparison of zonal avifaunas, such as was made by Todd and Carriker (1922: 59); but because a species does not necessarily occur throughout the zone to which it is attributed, it is desirable to compare the number of species found in small areas at different altitudes in the same region. At Finca "La Selva," in the Caribbean lowlands of northeastern Costa Rica, Slud (1960:142) identified, in the course of one year, 331 species of birds, of which 269 breed in Costa Rica. Some 5,000 feet higher, on the headwaters of the same Río Sarapiquí which flows past La Selva, lies Montaña Azul, where in a year I found 156 species, of which 132 breed in Costa Rica. Probably if Slud had worked the vicinity of Montaña Azul as intensively as he worked La Selva, he would have found somewhat more birds than I did, for to me making detailed studies of selected species took priority over censusing the avifauna; also much of the time I was engaged in botanical collecting. Yet I doubt whether, within the area I worked, the list of birds could have been increased by more than 10 or 15 per cent, which would still leave it far short of that for the lowlands over which Montaña Azul looked. Just across the continental divide from the sources of the Sarapiquí drainage is La Giralda, where between 6,500 and 8,000 feet I found, in four and a quarter months, only 86 species, of which 71 are resident in Costa Rica. But the avifauna here was somewhat reduced by extensive deforestation, hunting, and other forms of disturbance by man.

For the Pacific slope of southern Costa Rica, I am unable to present a comparison of the avifauna at different altitudes so convincing as that between La Selva and Montaña Azul. On our farm beside the Río Peña Blanca in the basin of El General, with an area of 100 hectares or one square kilometer and the slight altitudinal range of 2,400 to 2,600 feet, I have identified in the course of 24 years 272 species of birds, of which 49 are migrants from the north and the remaining 223 breed in Costa Rica—134 species with nests or fledglings have been found on the farm. Unfortunately, I have not kept records in such a form that I can now determine how many species were seen in a year or a single season. Some of these birds were noticed only once in many years; but I should hazard the guess that from March to the end of June a diligent observer might record a total of 225 species, of which about 35 would be migrants from the north and 190 would be residents. This may be compared with the total of 206 species—27 species of migrants and 179 of residents—found in the same interval at Cañas Gordas, 1,000-1,500 feet higher.

In Guatemala, where on the Sierra de Tecpam, between 7,000 and 10,000 feet, I found 89 resident species in 13 months, the situation is different. As Griscom (1932) has shown, the Subtropical Zone of Guatemala is smaller in area and much poorer in species than the Temperate Zone, which occupies the extensive Guatemalan *altos* and is the home of many species of north temperate origin, which doubtless passed southward across the Isthmus of Tehuantepec during a glacial period when the temperature of this region was cooler than at present.

Some of the reasons for the reduction in the number of species of birds at high altitudes are fairly obvious. In tropical America as a whole, the total land area diminishes with altitude, and the several highland areas are isolated from each other as the lowlands are not. The continental lowlands provide larger areas in which new species can originate, and after they arise they have better opportunities to extend their range, within the zone to which they are adapted, than do the highland species. Mountain masses are islands in the air, and as with islands in the sea, the smaller and more isolated they are, the more limited their avifauna tends to be. This is shown by Todd and Carriker's (1922:59) illuminating comparison of the percentage of species at high levels in the isolated Sierra Nevada de Santa Marta with that in the Colombian portion of the continent-long Andean chain. Moreover, the lowlands seem to offer a greater variety of habitats.

I do not know to what degree, if any, the smaller number of species at higher altitudes is compensated by a larger number of individuals per species. Some highland species are certainly very abundant; but my impression is that the density of the avian population of all species is smaller at high altitudes than in corresponding habitats in the lowlands. If this is true, the cause is doubtless to be sought, not in the direct influence of the lower temperatures on the homeothermal birds themselves so much as in their effect on the rates of growth and reproduction of the plants, insects, and other poikilothermal organisms on which the birds depend for food. If insects, for example, multiply more slowly, they cannot yield as much food for birds as they do in warmer regions where their life cycle is shorter. I know of no comparative studies of the rates of development of organisms at different altitudes, but an indication of how development is retarded by altitude is provided by the time maize takes to mature at various elevations. Here in the valley of El General, at 2,500 feet, we plant our corn in March and harvest it in August, five months later. On the Sierra de Tecpam, at 8,500 feet, the maize was planted in the dusty soil in March and grew slowly until the advent of the wet season provided enough water for more rapid development. By the beginning of November, the more retarded ears were still so tender that they were blighted by the first frosts. The crop was harvested in December, nine months after sowing. At the highest altitude at which maize would ripen on Volcán Irazú, slightly over 9,000 feet, I saw cornfields that were planted in January and harvested in the following January, a year later. Doubtless there is a corresponding retardation of development with altitude in many other poikilothermal organisms, and this, by decreasing the rate at which food can be supplied, reduces the density of the avian population which high altitudes can support. The whole subject deserves further study.

Acknowledgments

Many of the studies included in this book could not have been made without the cooperation of those who provided living accomodations, and often more inclusive hospitality, in places where lodgings are not easy to find. For such friendly help I am indebted to Axel Pira and Doña Leonor de Pira, at their estates "Chichavac" and "Santa Elena" on the Sierra de Tecpam; to Charles H. and Dorothea Lankester, at the hacienda "Las Cóncavas" near Cartago; to Rafael Angel Fernández Soto, Doña Consuelo Rothe de Fernández, and Rafael Angel Fernández Rothe, at their dairy farm "La Giralda" on the Barba massif; to Darryl and María Cole, at "Loma Linda" near Cañas Gordas; and to Bob and Catherine Wilson and John Ozanne, at Las Cruces near San Vito de Java. In 1963 I worked while holding an Elsie Binger Naumburg Fellowship for the study of Neotropical birds, from the Frank M. Chapman Memorial Fund of the American Museum of Natural History in New York; and in the following year my field work was supported by a research grant from the same fund. In these two years my wife Pamela was with me in the highlands and helpful in many ways.

Eugene Eisenmann's most useful *The Species of Middle American Birds* (Trans. Linn. Soc. New York, 7:1-128. 1955) greatly facilitated the selection of names for the birds treated in this book. I have followed his scientific nomenclature throughout, and with a few exceptions, I have also used his English names. The manuscript was accepted by the Nuttall Ornithological Club 7 February 1966.

To all these people and institutions I am most grateful.

Alexander F. Skutch

Finca "Los Cusingos"
San Isidro del General
Costa Rica

Family TROCHILIDAE

VIOLET SABREWING

Campylopterus hemileucurus

The Violet Sabrewing is a large, stout, easily recognized hummingbird, about five and a half inches in length, of which nearly an inch and a quarter is accounted for by its long, conspicuously curved, black bill. The plumage of the male is nearly everywhere bright metallic violet-blue below and largely of this color above, where the forehead and crown are dusky and the wing coverts, rump, and upper tail coverts are metallic green or bluish green. The rounded tail is bluish black, with broad white tips on the three outermost feathers on each side. In the female, the upper plumage is bright metallic green or bronzy green; the under parts are gray, spotted on the throat with metallic violet-blue or sometimes with a solid patch of this color. Her outer tail feathers are white-tipped, as in the male.

This strikingly colored hummingbird is found through the mountains from southern Mexico to western Panama. On the Pacific slope of southern Costa Rica, I have recorded it at altitudes ranging from 2,500 to 7,500 feet, but below 3,000 feet it appears to be only a transient which does not remain to breed. Its preferred habitat is the undergrowth of the wet mountain forests, especially near their edges, but it is occasionally found in lush thickets at a considerable distance from heavy forest. Although it is widespread, and likely to be encountered wherever in the more humid parts of its geographical and altitudinal range a little luxuriant vegetation remains, I have seldom found this hummingbird common. In the tall, mossy forests in the Cañas Gordas region of southern Costa Rica, between 3,500 and 4,000 feet above sea level, it was moderately abundant.

Like other hummingbirds, the Violet Sabrewing visits a variety of flowers. It is fond of heliconias or wild plantains, and it enters banana plantations to sip the nectar richly secreted by the long white flowers clustered beneath the deep red, fleshy, upturned bracts. One February, years ago, a female or young male sabrewing took possession of an extensive group of Scarlet Passion Flowers (*Passiflora vitifolia*) growing in tall second-growth woodland beside a pasture. In these large, spectacular flowers of complex structure, the floral nectaries are protected by a sleeve of thick tissue which prevents hummingbirds with bills of moderate length from reaching the nectar. The principal pollinators, here in El General, are the Long-tailed Hermits, although an occasional Green Hermit from higher altitudes shares the nectar with them. Little Hermits often visit the passion flowers but, unable to reach the floral nectaries, content themselves with sipping the fluid from the extra-floral nectaries on the bracts and gathering tiny insects attracted by the sugary secretion. The sabrewing who claimed the passion flowers drove away the more slender hermits, who could visit them only when the stout intruder was not guarding them. Although it dominated the hermits, the sabrewing was far more shy of me than they were, and would dart away at my slightest movement. Occasionally, however, I saw that it pushed its long bill down through the collar to reach the floral nectary, just as the Long-tailed and the Green hermits did (Skutch, 1952).

Voice and Courtship

The Violet Sabrewing is one of the many species of hummingbirds of which the males sing tirelessly—and usually tunelessly—to advertise their presence to the females. Sometimes the singing male sabrewing has no near neighbor, but perhaps more often he performs within hearing of several others of his kind. The location which the sabrewing selects for proclaiming his presence is usually amid low, dense vegetation, either within the forest or at a distance from it. Usually he perches no more than 15 feet above the ground.

The assembly to which I gave most attention was located at the western end of the Barba massif in the Cordillera Central of Costa Rica, at an altitude of 6,500 feet. Here the sabrewings performed amid the dense marginal shrubbery of the heavy forest that filled a deep ravine, below a steep hillside pasture. Four males were strung out along the forest's edge at intervals of about 50 feet. Usually they performed while perching well concealed amid the undergrowth, but sometimes one of them chose a more exposed perch, where he could be seen from the adjoining pasture above him. In a weak, unmelodious voice, these sabrewings sang interminably *tsee tsee, tuc see, tu wit see* . . . . The tempo was slow and the notes often squeaky. Unlike many other hummingbirds, which begin their vocal exercises at break of day, these started rather late, between seven and eight o'clock in the morning. After that, I could count on hearing them at any hour until past five o'clock in the evening, when they fell silent as the sun sank low.

The sabrewing who performed on a slightly exposed perch at the forest's edge would from time to time spread his tail fanwise for a brief instant, sending forth a transient flash of white. More rarely, he held his tail spread out for a second or two, while he vibrated it rapidly in a striking display. Occasional pursuits, when these hummingbirds chased each other through the undergrowth so madly that I wondered how they avoided striking fatally against the crowded twigs, were also accompanied by much flashing of the white outer tail feathers. I found this assembly already active on 5 March 1963, and the sabrewings were still performing here on 5 July of the same year.

In the same region, at least two sabrewings sang not far apart amid the dense undergrowth of tall, cane-like bamboo, well within the oak forest at an altitude of about 7,000 feet. At the end of June, 1935, a lone sabrewing sang persistently on a low perch amid bushy growth that covered a steep roadside bank on a coffee plantation, far from forest, at about 3,000 feet in the Department of Quezaltenango, on the once heavily forested Pacific slope of northwestern Guatemala. The sabrewings evidently maintain their singing assemblies through much of the year, for in mid-September of 1938 several were tirelessly proclaiming their presence in low woods with tangled undergrowth, at the top of the steep slope that fell off into the gorge of the Río Caliente, near Cartago, Costa Rica, at an altitude of 4,500 feet.

Rufous Sabrewings perform in much the same manner as Violet Sabrewings, but these plain-colored hummingbirds of drier country choose more exposed situations. Amid the coffee plantations about Lake Atitlán in Guatemala in late October of 1933, I found two who perched 10 or 12 feet up on exposed twigs, slowly and deliberately repeating a single note, a dry *chip*, with frequent breaks in their calling.

Nesting

The only nest of the Violet Sabrewing that I have seen was found in the forest near Cañas Gordas, at an altitude of about 3,800 feet, on 23 May 1964. The nest had been built on a slender, horizontal branch of a small tree that leaned over the middle of a narrow, steep-sided ravine. Its site was about 12 feet above the bottom of the ravine and almost level with the ground on either side. The structure was a deep open cup, covered with green moss that draped gracefully all around it. It was so bulky that, until with a mirror I saw the two white, elongated eggs, I surmised that it was the nest of a small flycatcher.

Seated amid the dense undergrowth that grew around the head of the ravine, about 50 feet from the nest, I watched through the morning of 24 May. I saw only the female sabrewing, but I heard no note from her. She sat on her eggs with her head bent slightly back and her long, curving, black bill tilted upward at an angle of about 45 degrees with the horizontal. Her head, tail, and long wings projected well above the nest's rim. Sometimes she sat with her body parallel to the twig that supported her nest, facing into the tree, but more often she incubated with her body transverse to the twig. Whatever position she took on arriving she usually preserved throughout her long session. Indeed, save for frequently blinking her eyes, she rarely moved perceptibly while she incubated, and I did not once see her touch her eggs with her bill. From 5:57 a.m., when she returned from her first excursion of the morning, until 11:31, the sequence of her sessions and recesses was as follows (recesses in italics): 11, *34*, 44, *40*,

13, *10*, 61, *26*, 85, *10*. The sabrewing covered her eggs for 64 per cent of the forenoon.

Like smaller hummingbirds, the sabrewing did not hop upon her nest's rim before taking flight, but flew directly from her eggs. However, instead of rising right up from the cup, as a lighter hummingbird might have done, she fluttered over the rim and flew off horizontally. Similarly, on returning she approached in horizontal flight and skimmed over the rim into the nest. Unlike every other hummingbird that I have watched incubate through a morning, this sabrewing brought nothing to add to her nest. It was already sufficiently bulky, and its form and composition evidently made it difficult to strengthen its attachment with fresh cobweb.

When I returned a few days later to continue my study of this nest, it had vanished. Probably it fell from its slender branch while heavily sodden with the torrential rain of the preceding afternoon and night, and the swollen rivulet, above which it had rested, carried it away.

GREEN VIOLET-EAR

Colibri thalassinus

The Green Violet-ear is a hummingbird of medium size, somewhat over four inches in length. In both sexes, the prevailing color is glittering metallic green. A band of dark violet-blue extends from the lores over the cheeks to the ear coverts, and there is a patch of the same color in the center of the chest. The tail is metallic bluish green, crossed by a broad subterminal band of blue-black, and the wings are largely dusky. The bill, of moderate length and nearly straight, is black. The foregoing description is of *C. thalassinus thalassinus* of Mexico and Guatemala, with which this account chiefly deals. The race found in Costa Rica and western Panama, *C. thalassinus cabanidis,* differs in lacking blue on the chest and lores.

The Green Violet-ear ranges from near the Tropic of Cancer in Mexico to Bolivia and Venezuela. Everywhere a bird of the highlands, in Guatemala it breeds from about 5,000 to 11,000 feet above sea level and in Costa Rica from 5,000 to 10,000 feet and possibly higher. In Venezuela, the species is found from about 3,000 to 10,000 feet above sea level (Phelps and Phelps, Jr., 1958:191).

Seasonal Movements

In the mountains surrounding the Valley of Mexico, the Green Violet-ear is present chiefly as a summer or breeding resident. The bulk of the population, including evidently all the females and yearlings, usually returns in the second half of July, but the arrival of the hummingbirds may be delayed by as much as a fortnight if the rainy season starts very late. On reaching their breeding ground, the females promptly begin to build their nests and rear their families. After raising at most a single brood, the females, the young, and some of the adult males leave between the beginning of October and early November. Other adult males remain in the region; but they, too, may disappear in February if the winter turns out to be exceptionally dry. There is evidence that in the latter part of the nineteenth century even the male violet-ears were absent from the region for eight months every year.

Wagner (1945), to whom we owe the foregoing observations, surmised that the violet-ears from the Valley of Mexico and its surroundings migrated southward across the Isthmus of Tehuantepec to Chiapas, Guatemala, and even beyond, and that they nested again while in the south. While this is not impossible, I am aware of no evidence that violet-ears ever cross the lowlands which stretch from coast to coast at the Isthmus of Tehuantepec. The species breeds in western Guatemala, and presumably also in neighboring Chiapas, at the same time that it does around the Valley of Mexico, but the breeding season is more extended. Probably the population that nests in the Valley of Mexico descends to lower altitudes after the breeding season, as do Green Violet-ears in Central America. The species has been found as far north as San Luís Potosí, in the Upper Humid Tropical Zone, in April (Friedmann *et al.*, 1950:164).

On the Sierra de Tecpam, in west-central Guatemala, Green Violet-ears were abundant, singing and nesting freely, in January of 1933. As the dry season advanced and the flowers on which the hummingbirds depend for nectar passed from bloom, their number decreased. I saw no violet-ear anywhere in the vicinity between the end of March and 21 August, when I found a single one on the plains at the base of the range, at about 7,100 feet above sea level. On 17 September, I saw a violet-ear halfway up the mountain, around 8,500 feet; but I found none near the top of the Sierra, where they had been most numerous, until late October. These observations suggest that, as the end of the wet season approached, the violet-ears gradually ascended the Sierra de Tecpam from lower altitudes. However, in 1934, when the rainy season in western Guatemala was much less stormy than in 1933, I found violet-ears singing profusely on the southern slopes of the Sierra Cuchumatanes, above Chiantla, on 15 August, and a month later they were doing the same on the broad plateau which tops the range, at nearly 11,000 feet. At Saloma, on the northern side of the same high ridge, Baepler (1962:144) first noticed Green Violet-ears on 2 August. They soon became common from 6,800 to 8,600 feet.

In the highlands of Costa Rica, above 5,500 feet, Green Violet-ears of another race make themselves conspicuous by their tireless singing from early September to about the end of March. At La Giralda in 1963, violet-ears which sang freely in February and March were not heard after 1 April. Soon the species became rare on the mountain slopes between 6,500 and 8,000 feet, where earlier it had been abundant. My last record of its presence here was made on 25 May, when I found one individual at 7,400 feet. Although I remained in the locality seven weeks longer, I saw no more violet-ears.

That these hummingbirds descend to lower altitudes at the end of the dry season is suggested by observations that I made years earlier on the Caribbean slope of Costa Rica. On 19 April 1941, I found at least two violet-ears at only 2,100 feet in the valley of the Río Pejivalle. Higher up the mountain slopes, around 3,000 feet, these hummingbirds were abundant about flowering hedges of *Stachytarpheta* in the second half of June; but during a sojourn from 28 August to 6 September of the same year in the same locality, I saw not a single one, although the hedges were still blooming profusely. Evidently the violet-ears had ascended to higher altitudes to breed. In the valley of El General on the Pacific side of the country, I have not found the Green Violet-ear below 2,900 feet, and at this altitude very rarely.

The differences in the timing of the seasonal movements of the violet-ear in various parts of its range seem explicable by differences in the length and intensity of the dry season. In the western highlands of Guatemala, where the dry season lasts from about mid-October to mid-May, the drought usually becomes very severe toward its end. In the higher parts of the Costa Rican mountains, the dry season is shorter and milder, hardly beginning before January and usually ending by early April; indeed, on slopes exposed to the northeast trade winds, a rainless season can hardly be said to occur. Here there is an abundance of flowers at all seasons, with possibly a minimum early in the wet season when many kinds of fruits are ripening.

Habitat and Food

Violet-ears avoid the sunless depths of heavy forest, where flowers are few. They prefer openings amid the forest, its edges, thinned woodlands, light copses, shady gardens and pastures, and even heavily cultivated districts where a few trees remain to provide singing perches and nest sites amid cornfields bright with flowering weeds. On the Sierra de Tecpam, I found some singing males but no nest in the zone of mixed woods of pines, oaks, and other broad-leafed trees below 9,000 feet; in the breeding season, the population of violet-ears on the mountain was concentrated near its summit, amid the forests of cypress (*Cupressus Benthamii*) above 9,000 feet. Here the hummingbirds established themselves in or around flowery openings rather than in the lofty, sombre forest itself. One such area supported so many hummingbirds of this and other species that I came to designate it "The Hummingbirds' Hillside." When I first visited it in November of 1930, the steep slope had been recently swept by fire following a lumbering operation. A few small living cypresses that had survived the conflagration stood scattered among the gaunt, charred trunks of those which had been spared by the woodsman's axe only to succumb to the blaze. Between the trees living and dead was a dense new growth of low, scrubby plants, including an abundance of the mint *Salvia cinnabarina* in full bloom. The slope faced eastward and was basking in the warmth of the bright morning sunshine, which had finally dispersed all the mists of the night. Gaily colored hummingbirds swarmed here, feasting on the spicy nectar in the long red corolla tubes of the salvia.

When I returned to "The Hummingbirds' Hillside" at the beginning of 1933, many of the dead cypresses had fallen, but enough remained standing to provide singing perches for the violet-ears. A multitude of young cypress trees, in dense stands higher than my head, had covered much of the steep slope; but scattered among them were numerous small, bushy openings where the scarlet salvia thrived, still blooming generously and offering its rich nectar to the birds. Here the numerous Green Violet-ears associated with Broad-tailed Hummingbirds, White-eared Hummingbirds, Wine-throated Hummingbirds, and Cinnamon-bellied Flower-piercers, all of which shared the nectar with them.

The many species of *Salvia* which add so much color to the Guatemalan highlands at the beginning of the dry season are favorite flowers of the violet-ears. Amid the cypress forests near the summit of the Sierra de Tecpam, they chiefly visited a species with red corolla tubes. But when, after five months of absence, the violet-ears returned to the Sierra in September, the red salvias had scarcely begun to blossom. However, the low *Salvia cacaliaefolia* was already freely displaying its splendid large blue flowers, and the violet-ears drew heavily on these for their nectar. Their wariness in my presence contrasted with the confidence shown by the White-eared Hummingbirds who shared the blue blossoms with them; but the Magnificent and the Amethyst-throated hummingbirds who also visited the blue salvia were still more timid. At this season, the violet-ears also fed from the pea-like scarlet blossoms of a bean (*Phaseolus* sp.) that climbed over stumps in the cornfields.

On the high plateau of the Sierra Cuchumatanes in September, the violet-ears who sang in the scattered pine and alder trees that grew on the slopes above the flowery alpine meadows seemed to depend largely on the stout purple-and-white flower heads of the abundant tall thistles of a new species that I found there (*Cirsium consociatum*). Broad-tailed, Magnificent, and White-eared hummingbirds shared this source of nectar with the violet-ears. In the Tablazo Mountains of central Costa Rica in the following September, I found a singing assembly of violet-ears in a copse of low trees beside a pasture, and these hummingbirds appeared to be sustained chiefly by nectar from the red-flowered *Cuphea infundibulum* that flourished in the grove. The violet-ears which in April and June I found between 2,000 and 3,000 feet in the Reventazón drainage of Costa Rica seemed to have been attracted to this low altitude by the violet flowers of the straggling verbenaceous shrub *Stachytarpheta* that had been planted in long hedgerows through the pastures. These abundant little flowers provided food for swarms of hummingbirds of seven species, including many of the curiously colored Brown Violet-ears, generally considered to be rare in Central America.

It appears that much of the time the violet-ears have some particular plant on which they chiefly depend for nectar, but the species of this principal source of food varies with the locality and, in the same locality, with the season. Like most hummingbirds, violet-ears visit many kinds of flowers in the course of a year. They vary their diet with small insects which they catch while darting about in the air.

Voice and Courtship

The singing male.—On my first visit to the Sierra de Tecpam in November of 1930, my host led me along a logging road that traversed the slope which I later knew as "The Hummingbirds' Hillside." The sun was shining brightly, and as we walked through the bushy growth where the salvia displayed its fragrant red blossoms, we heard from every side a metallic *k'chink chink k'chinky chink,* incessantly repeated, which I recognized as a hummingbird's attempt to sing. Since I did not know the species, I determined to have a good look at the songster, which seemed easy, for the tinkle arose on all sides. But an abundant bird is not always an easy bird to see. Whether the hummingbird perched high on the dead twig of a cypress tree or low among the bushes, almost hidden from view, he was not too absorbed in his tireless singing to be careful of his safety. In the high position, he offered me only a view at long distance through my binoculars and, against the bright background of the sky, he appeared quite black and lustreless. Stalk as carefully as I could, I was unable to surprise one of these hummingbirds performing on a lower perch. I would advance with all caution toward a bush from which the voice seemed to issue, only to find at last that the hummingbird had vanished, without my having seen him fly. It was not a question of a "wandering voice," for the voice remained quite stationary if I did not pursue it too closely; still, it seemed but a voice disembodied. To get an adequate view of the little hummer was a challenge to my birdcraft which I could not deny, and I spent the better part of two hours in futile attempts.

The approach of midday did not silence this untiring voice, as it does that of many birds, and the mountainside continued to resound with the tuneless tinkling. Finally, just before noon, I saw a dark-colored hummingbird probe some of the red mint blossoms, then dart with baffling speed to perch on a dead twig of a cypress tree, where he began the now familiar chant. The slope was so steep that the bird, although high above the ground and not far away, was below the level of my eyes, where his metallic colors showed to full advantage in the bright sunshine, against the dark background of the cypress trees on a distant mountainside. I wrote a detailed description from which I later identified the hummingbird as *Colibri thalassinus.*

When I returned to the Sierra de Tecpam at the beginning of 1933, I found the violet-ears calling as incessantly as on the November day when I first made their acquaintance. Throughout January I dwelt near them and spent much time watching them. The singing male violet-ears preferred an exposed dead twig, 15 to 40 feet above the ground. There were considerable stretches of mountainside with an abundance of salvia plants but without suitable trees, and the hummingbirds who claimed these areas were content, perforce, to sing from the bare bough of a low bush, not more than 6 feet above the ground. Occasionally a male violet-ear of a retiring disposition chose such a low singing perch even when higher, more conspicuous ones were available in his territory. Each male had his own particular headquarters, and although when singing he was not restricted to a single perch, the perches from which he habitually performed were not separated by more than 20 or 30 feet. Within this space were usually several suitable twigs which he might employ, and he frequently changed from one to another. In this small circuit he was to be found day after day, hour after hour, morning, noon, and evening, fair weather and foul, singing as though his life depended on it.

In the majority of hummingbirds, as in a number of species of birds of other families in which the male takes no interest in the nest, he does not go in search of a female but advertises his daily presence in one particular spot, where often his rivals surround him, so that each female who is about to lay her eggs may freely choose the father of her prospective offspring from among the eligible ones assembled in the vicinity. And so, during the breeding season, to proclaim his presence seemed to be the single object of the male violet-ear's existence. When white frost covered the ground at daybreak, when low-lying clouds bathed the mountain in their chill mist, when a cold drizzle fell—no matter how harsh the weather might become, once the return of the dry season set off the violet-ear's singing, no temporary reversion to wet season days could make him suspend it. With his spirit undaunted by cold and dampness, he continued to sing when the dense cloud-mist that enveloped the mountain made him invisible at no great distance, and from all around me came the reiterated notes of unseen hummingbirds.

Although at first it had cost me so much effort to locate the sources of this small chatter which pervaded all the mountainside, after I had become familiar with the habits of the violet-ears, I had not difficulty finding them, or examining their glittering plumage at my leisure.

When I had located the singing perches of any particular individual, I had only to sit quietly in view of them, certain that in a few minutes the bird would return to resume his interrupted monologue.

The voices of the violet-ears were too thin and sharp to be melodious, but their song had at least a suggestion of rhythm. At times I believed that I could detect a set phrase of six or seven squeaky syllables, but at other times I failed to recognize any structure in the utterance. These hummingbirds began to sing in the frosty dawn, as soon as the light became strong enough for objects and colors to be clearly seen, and they continued throughout the day, pausing only to sip nectar from their salvia blossoms, to court a female, or to drive an intruder from their territory. Since, at least on "The Hummingbirds' Hillside," the territory of each singing male supported close at hand sufficient flowers to supply the food he needed, he was never obliged to go far off to forage, and he reduced to a minimum the time he devoted to his meals.

One morning at the end of January, I made a careful record of all the activities of a male violet-ear from 9:52 to 11:15 a.m. Of the 83 elapsed minutes, he spent over 70 singing on his perches. There were 13 intervals of practically continuous singing, ranging from less than one to 11 minutes in length. His longest absence from his singing perches was 3 minutes, when he chased a trespassing violet-ear, or perhaps engaged in courtship exercises with a female. Two other chases of intruding violet-ears interrupted his singing for only 10 seconds on each occasion, and when he pursued a trespassing Wine-throated Hummingbird he was away for about the same interval. On another occasion he was inexplicably out of sight for 2 minutes. This violet-ear's periods of feeding lasted about a minute and often much less. During the 83 minutes of my record, he spent about 7 minutes visiting the salvia blossoms in my field of vision; without a stop-watch, I could not make a more accurate determination. Evidently his only other nourishment during this period consisted of the insects, too small for me to see, which he caught on aerial darts from his singing perches that hardly interrupted his singing.

Likewise the violet-ear's occasional shifts from perch to perch within his territory scarcely broke the flow of his squeaky notes. While singing, he continually turned his head from side to side, and from time to time he spread his wings and vibrated them so rapidly that they dissolved into a haze. At the end of each spell of singing, he simply dropped down to the flower beds beneath his perches and sipped the abundant nectar from many of the red corollas during the minute or less that he devoted to his meal. Even in the course of such a hastily snatched repast, he could not refrain from uttering a few notes of his song. Sometimes while hovering before the flowers he made a loud, rapid clicking, apparently with his wings. His hunger and thirst assuaged, he shot directly upward to one of his singing perches with his forehead covered with pale yellow pollen and resumed his chant the moment his feet touched the twig; sometimes he began even before he alighted.

And so the violet-ear continued tirelessly to pour forth his notes, until the landscape grew dim in the evening twilight, and the nocturnal chill began to descend upon the mountain. Then he flew down to suck the nectar from his salvia blossoms; and I noticed that his neighbors, who a few minutes before had maintained an intermittent twitter, had also become silent. There was still sufficient light to distinguish colors and minute objects, and the Rufous-collared Thrushes were chattering in the tops of the cypress trees. The violet-ear, after hovering in front of a few flowers, settled on a low perch and preened himself for several minutes. Then he made a much longer round of his flower beds than he did during the day, to fortify himself against the frosty night air by a liberal supper. At the conclusion of his meal, he darted away over the bushes and vanished, to pass the night where I could never find him. In late January, he started to call at 6:25 a.m. and continued until 6:12 p.m. The number of notes which he had uttered in nearly 12 hours, had anyone the patience to count them, would exceed belief. At the height of the breeding season, the male violet-ear seems to be above all a highly efficient machine for transmuting nectar into squeaks.

Relations of neighboring males.—Many of those who have written about hummingbirds expatiate on their pugnacity, until one who knew them only from books might conclude that they are fierce little animals who delight above all in maltreating each other. Although I have watched many kinds of hummingbirds under varied conditions, I have scarcely ever seen one inflict even slight injury on another. It is true that when a number of individuals come together, as in a flower garden or about the crown of some flowering forest tree, one often dashes at another as though to transfix it with his rapier-like bill; but the bird attacked nearly always flees in time to avoid the impact, and the assault leads only to an aerial pursuit of astonishing swiftness and a harmless outcome. The spectacular chases in which these sprightly little creatures so frequently engage are no more proof that they are the most pugnacious of birds than that they are the most playful of birds.

On "The Hummingbirds' Hillside" each male violet-ear was usually in sight and hearing of three or four others of his kind as they sang on their regular perches. But he left his neighbors alone, and they left him in peace, so long as the domain of each was not invaded. The singing perches of each bird seemed to be respected by all, and I never saw a violet-ear molested while he sang in his chosen position. Sometimes, however, the flower beds which belonged to one individual would be visited by one of his neighbors. Whenever a violet-ear saw a trespasser hovering among the flowers that he claimed as his own, he never lost a second in darting toward the intruder, who admitted his guilt by immediate flight, when pursued and pursuer became mere dark streaks over the green slope. The holder of the territory was usually back on his singing perches in such short order that there was no time for him to have done more than chase the intruder well beyond the confines of his territory. Usually other kinds of hummingbirds, including Broad-tails and White-ears, were not molested while they visited the flowers claimed by a violet-ear; but occasionally they, too, were driven away.

The singing perches of the violet-ears were never so close together as, for example, those of hermit hummingbirds (*Phaethornis* spp.), whose singing assemblies are in the dense thickets or the tangled undergrowth of the forest in the lowlands, not yet so close as those of the White-eared Hummingbirds among the oak woods lower on the Sierra de Tecpam. The spacing of hummingbirds in their singing assemblies is evidently related to the carrying power of their voices, which in turn is determined not only by their actual loudness but also by the environment. Since the sharp metallic note of the violet-ears carries far over the open mountainside, they need not be close together to be within hearing of their nearest neighbors. The headquarters of the several violet-ears on "The Hummingbirds' Hillside" were from 50 to 100 yards apart, depending on the character of the terrain and the location of good perches.

Relations of the sexes.—While it was easy to see and to hear how the male violet-ear attempted to attract the other sex, what he did when a female responded to his tireless solicitation was more difficult to learn. In the first place, the sexes of this hummingbird can hardly be distinguished by appearance, so that my only warrant for ascribing the male sex to the persistent songsters is analogy with other species of hummingbirds (and birds of other families) with similar modes of courtship and pronounced sexual differences in plumage. I am not sure that, when a male pursued another violet-ear which had been sipping nectar from his blossoms, it was always a case of driving away a trespasser. Perhaps at times the second bird was a female, and the flight a nuptial pursuit rather than the repulsion of an invasion. When hummingbirds chase each other, the human eye can hardly follow. Once, however, I watched two violet-ears amicably feed close together. They were evidently a male and a female, for at that time every adult male seemed to have his own territory which he defended with zeal. Finally, while one of the two was sucking nectar, the other perched on a bare twig in front of her and quivered his spread wings. The supposed female showed no appreciation of this display, which was of momentary duration. Perhaps because of the lateness of the season, I failed to see a male and a female fly side by side as Wagner (1945:171) has described.

Duration of singing in Guatemala.—By mid-February, the *Salvia cinnabarina* felt the effects of the long continued dry weather and the heavy nocturnal frosts, and there was a rapid decrease in the abundance of its flowers. With the passing of the blossoms that were their chief source of food, the violet-ears rapidly fell silent. By 10 February, five males that I knew well were missing from their posts; but others sang on. When I returned to "The Hummingbirds' Hillside" on 17 February, desolation reigned there, a blight seemed to have fallen over the slope. The silence which prevailed during the middle of the day, after the violet-ears ceased to sing, was accentuated by the absence of a chorus of insect voices such as one hears at lower altitudes; there was no chirping of crickets, sizzing of cicadas, or rustling of grasshoppers. The insects present on the mountaintop at this season were nearly all small, and if they produced sounds, these were pitched too high for me to hear. During the months when the hummingbirds call on the mountaintops, their voices, by virtue of their quality, persistence, and number, occupy the place in the system of audible nature which in the tropical lowlands, and in summer in the temperate zones, is filled by those of insects.

Soon after they became silent, the violet-ears began to withdraw from the Sierra de Tecpam. By mid-March, when the salvia blossoms had all but disappeared, the decrease in the number of these hummingbirds was very evident. After March, I saw no violet-ear for nearly five months. Then, late in August, while wandering over the plains near Tecpam, I watched a lone violet-ear drinking nectar from the scarlet blossoms of a wild bean (*Phaseolus* sp.) which had clambered over a shrub beside a stream, converting it into a veritable burning bush. The gradual reappearance of the violet-ears at points successively higher on the Sierra de Tecpam has already been described.

With the advent of October, and the prospect of sunny days soon to be, the red salvias put forth a few tentative blossoms, which increased slowly in number as the month advanced. By the middle of October, the usual date of the beginning of the dry season, I heard the familiar singing of the violet-ears for the first time in nearly eight months. I had come upon a singing assembly on a hillside covered with tall raijón bushes (*Baccharis vaccinioides*), among which stood tall alder trees (*Alnus arguta*), at a point about midway between the base of the Sierra and its summit. This was two months later than I heard violet-ears performing on the Sierra Cuchumatanes in the following year, as already told (p. 123). Soon the violet-ears were singing quite generally all over the highlands. At the end of October, I walked from the summit of the Sierra de Tecpam to Lake Atitlán, a drop of 5,000 feet. All along the path, wherever there were suitable open woods or scattered trees for perches, I heard their reiterated tinkle. There was now no dearth of flowers to supply them with the essential nectar, for everywhere the trailside was gay with myriad bright blossoms. A few violet-ears even sang among the shade trees of the coffee groves on the flood plains beside the lake, at an altitude of only 5,000 feet; but here, where subtropical conditions prevailed, they were far less numerous than on the steep escarpments above the shores, which rise rapidly into the Temperate Zone.

When I returned to the Sierra de Tecpam at the beginning of November, I found the violet-ears singing tirelessly on "The Hummingbirds' Hillside." One violet-ear, whose song was punctuated by a single clear, ringing note which served to distinguish him from all his neighbors, performed on the same perches where I had watched him ten months before; the peculiarity of his voice assured me that he was the same bird. Other violet-ears sang in the same trees which had been used for singing perches during the preceding January, and most were probably the same individuals, but they possessed no distinguishing peculiarities by which I could recognize them. In Mexico, a violet-ear with a peculiar call occupied the same isolated tree during the breeding seasons of four consecutive years (Wagner, 1945:168).

Duration of singing in Costa Rica.—In the Costa Rican highlands, where the dry season starts much later than in western Guatemala and is less prolonged and severe, the period of singing of the Green Violet-ear is much longer than I found it to be on the Sierra de Tecpam. At an altitude of 5,700 feet in the Tablazo Mountains at the southern edge of the Central Plateau, I found a well-established courtship assembly as early as 10 September 1935, at the height of the wet season. In a copse of low trees beside a pasture, three violet-ears performed on perches from 4 to 15 feet above the ground and 60 to 90 feet from each other. On the following 14 March, when I walked for miles along the crest of the Cordillera de Talamanca at an altitude of nearly 10,000 feet, I heard violet-ears singing tirelessly all along the rough trail, except where the forest was tall and dense. On 7 November 1964, violet-ears were in full song on the southern slopes of Volcán Barba, between 6,000 and 7,500 feet; and in the same region I found them proclaiming themselves with unabated zeal on the following 8 February. Over vast stretches of territory in the highlands of tropical America, from Mexico far south into the Andes, the squeaky song of the Violet-ear is one of the most prevalent of natural sounds through a large part of the year.

Nest

In Guatemala in 1933, I found four nests of the Green Violet-ear, and in Costa Rica in later years I saw four more. All the Guatemalan nests were on "The Hummingbirds' Hillside," where the earliest was discovered with recently laid eggs on 10 December. The other three had been found here in the preceding January and February, when the breeding season was further advanced.

All four of these nests were built on horizontal lower branches of cypress saplings, where they were supported by lateral twigs. Three of them were at approximately three feet above the ground, the fourth 5½ feet up. Each was beside a logging road or a small clear space rather than in the midst of the dense stands of young cypresses, where they would have been better concealed but the parent's flying approach and departure would have been impeded. Each nest was a broad, rather shallow cup, composed mainly of mosses thickly felted, which in some cases remained green but in others had dried and turned brown. Cobweb bound the mosses together and served for the attachment of the many small leaves, dried and shrivelled, that "decorated" the outside. To the outer surface of one of these nests were fastened some long shreds of the fibrous inner bark of the cypress and some twiglets from the same tree, in addition to a few shrivelled dead leaves. The use of dry leaves and the like rather than lichens for covering the outside of the nest is distinctive of the Green Violet-ear, at least in the northern part of its range. Some of the nests were lined with the downy pappus of composite seeds and others with downy feathers, including the red feathers of the Pink-headed Warbler. These nests measured 2⅛ to 2½ inches in over-all diameter by 1⅛ to 1½ inches in height. The internal diameter varied from 1¼ to 1½ inches even in the same nest, and the depth was from ¾ to one inch.

The four Costa Rican nests, all found within about eight miles on the Cordillera Central, were more diverse in situation and structure than the Guatemalan nests. The earliest, found with two eggs on 9 October 1937, was attached to the upper side of the stipe of a fern which hung from a roadside bank, about four feet above the muddy Sarapiquí trail that was used by many horsemen and pedestrians. This nest, at an altitude of 5,300 feet, was the lowest above sea level that I have seen. It was composed almost wholly of the ramenta or scales from the stipes of ferns, of a very light color. A few small, inconspicuous bits of green moss were attached to the outside, and some long pieces of dead leaf hung untidily below it. On the following 9 March, a newly completed nest was found 100 feet higher in the same neighborhood. This nest was attached near the apex of a slender herbaceous stem hanging over a bank, beside a path that ran between a pasture and woodland. The bank was here about 6 feet high; the nest was 5 feet above its base and 8 inches out from it. The open cup was loosely constructed of decaying bits of grass blades and moss, and sparsely lined with down.

Twenty-five years later I was shown a nest at about 7,700 feet at La Giralda. This structure was attached, 55 inches above the ground, to a slender vine that dangled below a massive, irregular trunk that leaned over a forest path. The bulky cup was covered with green and brownish moss and lined with light-colored seed down. It measured 3 inches in diameter by 2½ inches high, or 10 inches including a tuft of moss that dangled far below the bottom. Inside it was 1¼ inches in diameter by one inch deep.

The fourth Costa Rican nest was on the same mountain slope, but 1,300 feet lower. It was fastened to some slender roots dangling beneath the overhanging top of a 12-foot-high cut bank beside a paved highway, over which much motor traffic passed. The overhang of roots with attached soil not only formed a solid roof above the nest, protecting it from rain, but it draped far down on the outer side, forming a broad, opaque screen between the nest and the road. The nest was, in effect, in a little cave open below and at both ends. The mosses of which it was largely built had turned brown, because they were continuously dry.

It was noteworthy that all the four Costa Rican nests, built by *C. t. cabanidis,* were attached to downward-drooping stems or dangling roots or vines, whereas the four Guatemalan nests, made by the nominate race, were on horizontal limbs of cypress saplings. Likewise, six of the eight nests of this race found by Wagner (1945:172) were on forks of thin branches of small trees, in this case oaks, and the other two were in vertical forks of the suffrutescent mint *Salvia polystachya.* These nests were composed of moss, sometimes with a lining of vegetable down, and covered on the outside with wide dry grass blades, which sometimes hung as much as 8 inches below the bottom. The few available records suggest a consistent difference in the choice of nest sites by *cabanidis* and *thalassinus.* The latter appears more regularly to build its nests of moss and to attach many shrivelled dicotyledonous leaves, grass blades, or similar materials to the outside. Both races prefer low sites. Wagner's Mexican nests ranged from about 16 inches to 6 feet above the ground. With the exception of the nest within the cave-like overhang at the top of the roadside bank, all that I saw in Guatemala and Costa Rica came within this range of heights.

Eggs

Each of my nests contained two eggs or nestlings, the almost invariable brood size of hummingbirds. The pure white eggs had the narrow, elongate form typical of the family, and those in three nests in Guatemala measured 13.5 by 8.7 and 13.1 by 8.7; 13.5 by 9.1 and 13.9 by 9.1; 13.5 by 8.7 and 13.9 by 8.7 mm.

In the earliest of the four nests found on the Sierra de Tecpam, eggs were laid about 7 December 1933. The latest nest contained, on 9 February 1933, two eggs which had vanished by 23 February. Probably in other parts of Guatemala, or in other years, the nesting season begins earlier and is more prolonged than these dates indicate. Baepler (1962:144) reported that, at Saloma, Green Violet-ears nested in August and September, but he did not reveal on what evidence this statement was based. In the Valley of Mexico, the violet-ear nests from July to October and has time to raise only a single brood (Wagner, 1945).

In Costa Rica, my earliest record for eggs is 9 October 1937. In another nest in the same locality, the second egg was laid on 10 March 1938; and at higher altitudes in the same region, a nest with recently hatched nestlings was found as late as 21 March 1963. At 9,700 feet on the Cordillera de Talamanca, I found, on 14 March 1936, a fledgling that had apparently left its nest only a day or two earlier. In Costa Rica, the breeding season is long enough to permit the rearing of two or possibly three broods. It is noteworthy that no nest has been recorded from any part of Central America in April, May, or June, when the nesting of the avian population as a whole is at its height.

Nestlings

Care of young nestlings.—At their nests, Green Violet-ears are more timid than many other kinds of hummingbirds. Even when I used a blind, one of the violet-ears on "The Hummingbirds' Hillside" refused to build in my presence, and an attempt to watch incubation at a neighboring nest resulted in the desertion of the eggs. At a Costa Rican nest, both eggs hatched 16 days after the second was laid. The incubation period at one nest in Mexico was approximately the same (Wagner, 1945: 177).

The nestlings, hatched in a rudimentary state of development, had black skin with a few tufts of tawny down along the middle of the back. Their eyes were tightly closed and their very short bills were yellow. Now the parent's attachment to her nest grew stronger, and she was not kept away by a blind placed close to it. I passed a total of 8 hours—five in the forenoon and three in the afternoon—watching a female attend two nestlings from about five to seven days old, still naked and sightless. When I arrived at seven o'clock on the morning of 27 January, the mountainside was white with frost wherever the ground was not shielded by trees from radiation into the clear night sky. By nine o'clock the brilliant sunshine, which had melted all the frost on which it fell, beat down upon the exposed nest, causing the nestlings to open their mouths and pant so violently that they made a clicking sound, evidently by the opening and closing of their glottis. Their mother gave them little protection from either the early morning chill or the strong insolation of mid-morning. In the 4 hours from 7:05 to 11:05 a.m. she brooded them eight times, for intervals ranging from a few seconds to 24 minutes. But only one session of brooding exceeded 8 minutes, and her total time on the nest was only 48 minutes, or 20 per cent of the 4 hours. Once she left the nestlings exposed for 128 minutes continuously. She fed them nine times.

In the afternoon, this eastward-facing slope was in the shade, and the air soon became so chilly that, despite the coat and vest that I wore and the measure of protection that the cloth blind afforded, I was uncomfortably cool. Still the nestlings, scarcely covered by their sprouting pinfeathers, were brooded very little. In the 3 hours from 2:55 to 5:55 p.m. on 28 January, their mother covered them for six intervals, ranging from 4 to 9 minutes and totaling 40 minutes, or about 22 per cent of the elapsed time. She fed them nine times.

In the chill air early on the following morning, the parent violet-ear brooded only once, for 10 minutes, in the hour from 6:44 to 7:44. She fed the nestlings three times. She always approached her nest through the thicket above it on the slope rather than through the clear space beside it, thereby making her arrival less conspicuous. Usually she brooded facing into the thicket.

As I had earlier found with lowland species, young hummingbirds are amazingly hardy. These nestling violet-ears, while still practically naked, endured exposure to cold air and, more surprisingly, intense radiation for periods that might have been fatal to even feathered nestlings of many larger birds.

In the total of 8 hours, the two nestling hummingbirds were fed a total of 21 times, which corresponds closely to the rate of feeding of Scaly-breasted Hummingbirds of about the same age (Skutch, 1964a) and differs little from that of the White-crested Coquette (Skutch, 1961). The nestling violet-ear's food, which I could not see, seemed to consist in part of nectar from the salvia blossoms and in part of the minute insects which the parent often caught as she darted erratically back and forth in the air above the nest.

Sanitation of the nest.—After feeding the nestlings, the mother sometimes, especially in the early morning, remained standing on the nest's rim while she reached down into the bowl, picked up their accumulated droppings one by one in the tip of her bill, and tossed them out by sideward jerks of her head. Later I watched White-eared Hummingbirds with tiny nestlings clean their nests in the same way. It is sometimes said that young hummingbirds themselves keep their nest clean by rising up to eject their excreta over its rim; but they can do this only after they are considerably bigger and stronger than when they hatch, although before their eyes open and their feathers grow out. For the first few days, the droppings are voided inside the nest; then there is an interval when the young are strong enough to eject them on, but not beyond, the rim. And while the mother broods the nestlings at night, they are of course unable to rise up and shoot their excreta from the nest. The parent removes the droppings which remain inside the nest but leaves those that litter the rim. Her sanitary practices are far less perfect than those of nearly all passerine birds, which remove all droppings as soon as they are voided rather than after a number have accumulated, as the hummingbird does; which remove them not only from all parts of the nest, but even from surrounding foliage; and which, instead of merely throwing the waste matter aside, carry it off to a distance. Passerine birds find it easier to keep the nest clean because their nestlings' excreta are enclosed in gelatinous sacs, while those of hummingbirds are not.

The parent's behavior when the nest was covered.—While the nestling violet-ears were panting most violently in the hot morning sunshine, I placed a small green leaf over the nest to give them some relief. When their mother returned and found the leaf covering the top of her mossy cup, she circled around it with vibrating wings and spread tail. After less than a minute, she alighted on the rim and pushed her bill under the near edge of the leaf, only to withdraw it at once and stick out her long white tongue. She repeated this performance several times, trying to feed her hidden nestlings, but evidently without success. Then she edged onto the nest, pushing the leaf to one side with her body, and sat in the nest partly on the leaf. After a while she rotated in the nest, thereby pushing the leaf a little farther outward. Soon she turned again and sat with her bill above the leaf, which was now folded over the rim. After brooding for 24 minutes, partly on the nestlings and partly on the leaf, she darted away when a sudden gust of wind rustled the blind. She had made no deliberate effort to remove the obstructing leaf.

After I had finished the foregoing observations on feeding and brooding, I again covered the nest with a small green leaf. In the 35 minutes that I left it there, the hummingbird returned repeatedly, to stand on the rim and lower her bill, as though trying to feed the nestlings that she could not see. She also made several attempts to brood, sitting on the leaf, which separated her from the nestlings, but she could not make herself comfortable and soon departed. Once, while she stood on the rim, a nestling suddenly rose up, moving the leaf and frightening her away. When convinced that the parent hummingbird was unable to solve the problem of reaching her nestlings, I removed the obstruction. I am sure that the leaf was not too heavy for her to move if she had tried. In passerine birds, I have noticed that the parents are more likely to remove a white obstruction, such as a handkerchief or a piece of paper, than a green leaf that covers their nest. White is the color of the fecal sacs that they are in the habit of carrying away (see Skutch, 1950b:224; 1960:108-109).

Care and behavior of older nestlings.—The growing bodies of the young violet-ears stretched the walls of their nest, then burst it asunder like an outgrown garment. The rent extended down one side of the nest and across its bottom. When I noticed that this had occurred, the nestlings were about 16 days old but still unable to fly, so that they were obliged to remain about a week on the flattened remains of their nest or else resting on the cypress branch beside it. Their mother ceased to brood them, although the nights were very cold and usually there was a heavy frost. The young hummingbirds, who were by this time well clad in plumage, slept with their feathers all standing on end, which probably had the effect of creating a thicker layer of still air around them and providing better insulation for their little bodies.

I watched these nestlings again on the morning of 9 February, when a dense mantle of cloud covered the mountaintop and a cold, driving rain, such as on these heights occasionally occurs even in the midst of the dry season, fell through much of the forenoon. Between 7:35 and 9:13 a.m. their mother came to feed them nine times. Usually on a single visit she fed each nestling once, but on one visit she regurgitated twice to each one, and on two visits she gave food to only one of them. To transfer the food, which was never visible when she arrived, she inserted her long bill deep into a nestling's throat and pumped up the nourishment by muscular action, in the usual manner of hummingbirds. Once she found one the nestlings perching beside rather than on the remains of the nest, facing away from her. To receive its meal, this nestling twisted back its head until it was nearly upside down, at the same time opening its mouth to admit its mother's bill. Even with the nestling in this strained posture, the transfer of food was accomplished, for when the parent had finished regurgitating, there was a big lump on the side of the young bird's neck, where its crop was situated.

With liberal meals of salvia nectar and small insects, the nestling violet-ears throve despite exposure to the cold, thin night air and occasional dark, drizzly days. Even on the morning of chilling, driving rain when I watched them, they showed no sign of suffering. On the contrary, they demonstrated their good spirits by devoting much time to preening their new plumage, and by frequently raising their wings above their backs and beating them into a haze, while clinging to the remains of the nest to prevent being lifted into the air by these vigorous exercises. They had learned to recognize the whirr of their mother's wings, and when they heard her approaching they looked expectantly around, sticking out their long, white tongues in anticipation of the forthcoming meal, just as the parent stuck out her own tongue after she had delivered it. On the following day, when the sun shone again and I had set my camera on its tripod before the nest to photograph the act of feeding, one of the nestlings uttered a single *peek* while its mother, suspicious of the strange object, delayed to approach them. Otherwise, the young were quite silent.

At the age of about 19 days, these nestlings had green upper plumage, but the feathers of the crown, hindneck, and back were margined with brown, and natal down still adhered to the tips of some of those on the rump. The tail, still very short, was metallic blue-green, and the remiges were dull black. The under parts were grayish green, becoming light gray on the flanks. The nestlings lacked the blue or violet which covers the center of the chest of this race of the Green

Violet-ear, but they already showed some blue feathers on the sides of their heads, especially in the auricular region. They had black bills and dark brown eyes.

Since I did not know the exact age of these nestlings, I could not determine the exact length of their nestling period, which was 20 days or more. In Mexico, Wagner (1945) found that Green Violet-ears normally leave the nest when 23 to 25 days old, but if they are raised in very inclement weather, their development may be so retarded that they remain in the nest for 28 days or more. I found a similar range in the length of the nestling period of the Scaly-breasted Hummingbird, even at altitudes much lower than those at which the violet-ear breeds (Skutch, 1964a).

The male and the nest.—In other species of hummingbirds in which the males gather in courtship assemblies, the nests that I have found have nearly always been beyond sight and hearing of the singing males. This was also true of the violet-ears' nests that I found in Costa Rica. On "The Hummingbirds' Hillside," however, the nests of the violet-ears that I saw were rather distant from the males' singing posts but not beyond hearing of their voices. One nest was about 125 feet from the nearest singing station. Two singing males were established at about 150 feet, in different directions, from the nest to which I gave the most attention.

Since the sexes of the violet-ear are so similar in appearance, I should have known that a male was attending a nest only if he came to incubate, brood, or feed the nestlings while the other parent, who had done one of these things, was still in view; but this never occurred. Once, indeed, while the female brooded the nestlings, another hummingbird of her kind flew up and hovered before the nest. In a moment it flew off, while the parent continued to sit with no sign of hostility. Further evidence of the male violet-ear's lack of attention to his offspring was provided by his almost constant presence on his singing perches. Wagner (1945) likewise failed to find a male Green Violet-ear attending a nest.

Years after my observations were made, Moore (1947) published a record of incubation by a male Gould's Violet-ear in Ecuador; and later, in Venezuela, Schäfer (1954) found a nest of this species at which the male incubated and afterward fed the young. This male was readily distinguished from the female, who also attended the nest, by his abnormal coloration. Since, as I saw in Ecuador and others have reported, the males of Gould's Violet-ear sing quite persistently, much in the manner of their smaller relatives the Green Violet-ears, one wonders how they combine this activity with their parental ministrations—if, indeed, they regularly participate in the latter.

BLACK-BELLIED HUMMINGBIRD

Eupherusa nigriventris

A tiny hummingbird slightly over three inches long, the Black-bellied is unmistakable in its striking attire of green, black, and white. In the male, the forehead, crown, and all the under parts are velvety black, with the exception of the white under tail coverts. The upper plumage, posterior to the crown, is metallic bronze-green. The four middle tail feathers are blackish glossed with bronze, and the three lateral ones on each side are white, sometimes with dusky tips. The female has the whole upper parts metallic bronze–green. Her four middle tail feathers are dull greenish bronze; the two outer ones on each side are white; and the third from the outside is mostly white, with dark edges. Her under parts are dull grayish white or pale brownish gray. In both sexes, the bill, nearly straight and of moderate length, is largely black.

This little hummingbird is known only from the Caribbean slope of southern Central America, from the Cordillera Central of Costa Rica to the Volcán Chiriquí in Panama. Within this restricted range, it lives from about 2,000 to 7,000 feet above sea level. In 1937 and 1938, I found it abundant at Montaña Azul between 5,000 and 5,500 feet. Here I saw it most commonly about the edges of the wet forest and in adjoining clearings and bushy growth, where the shrubs, vines, and epiphytes flowered more profusely than within the dark woodland. Aside from its unusual coloration, this hummingbird lacked a pronounced character; and I have nothing noteworthy to record on its habits. But I had the good fortune to find two of its nests, which seem never to have been described.

Nesting

The first nest was 6½ feet above the ground in a shrubby *Siparuna*, beside a narrow road that traversed a strip of low, tangled second-growth close by the forest, at an altitude of 5,400 feet. When found on 31 July 1937, the tiny cup appeared nearly finished, but the female was still adding to it. Her tameness contrasted strongly with the wariness of the Amethyst-throated Hummingbird, the Green Violet-ear, and other members of the family that I had studied in the Guatemalan mountains. While two of us sat on a log across the road, only 7 feet from the nest, the Black-bellied Hummingbird continued to build as though we did not exist. While she sat shaping her nest, I could approach to within a yard without disturbing her. Her little chalice appeared to consist chiefly of brown scales (ramenta) from the fronds of large ferns, and it was sparingly decorated on the outside with bits of moss and lichens. Much more moss was attached to the exterior after the eggs were laid. The nest was well sheltered from the frequent rains of this season by the broad leaves of the supporting shrub.

Two minute white eggs were laid in this nest between 4 and 7 August. On a number of visits, I found only the female incubating. She would remain sitting until I almost touched her, then fly to a low twig across the roadway and stay there until I approached within arm's length. Then she retired to another perch a few feet farther away and permitted me to come as close as before. When her eggs were on the point of hatching, she for the first time permitted me to touch her tail. Then she hovered on the opposite side of her nest, facing me, and spread her tail so that I could distinguish the three white feathers on either side. After this pretty display, she settled on a low perch a few yards away. Between 23 and 27 August her eggs hatched, after no less than 16 days of incubation. A few days later, the nestlings vanished.

The second nest was situated in the forest, far out on a horizontal branch of a shrubby epiphytic heath (*Cavendishia* sp.), at about 12 feet above the ground and 5,300 feet above sea level. When found on 1 March 1938, it contained two naked nestlings. Supposing that this female would be as confiding as the first, I watched at a distance of several yards; but she refused to feed her nestlings in my presence. Instead, she flew around repeating a sharp *cheep,* and once approached to examine the intruder while she hovered within arm's length of his face. I went farther off to watch from behind a screen of foliage; and soon the hummingbird, evidently failing to notice me, flew up and alighted on the nest to feed the nestlings. But before she started to regurgitate she spied me and flew directly up, to hover for a few seconds at arm's length before my face, as before. After this close scrutiny, she flew off in the opposite direction, to perch on an exposed twig and await my departure. Evidently this hummingbird had little fear of me but tried not to reveal the location of her young. The close approach that hummingbirds, especially those that dwell in the dimly lighted undergrowth of the forest, frequently make to scrutinize an intruding man, suggests that they are nearsighted, which would not be surprising, considering at what close quarters they must operate to catch minute darting insects in their bills.

AMETHYST-THROATED HUMMINGBIRD

Lampornis amethystinus

The Amethyst-throated is a fairly large hummingbird about four and a half inches in length. In the male, the upper parts are chiefly metallic bronze-green. The remiges are dusky and the tail is black, with brownish gray tips on the outer feathers. A broad dusky band extends from the base of the bill over the cheeks and ears and down the sides of the neck. This is bordered above, from the eye backward, by a narrow but conspicuous white stripe, and below, on the cheeks, by a less prominent narrow streak of cinnamon-buff. The throat is bright metallic reddish purple; the breast and abdomen are brownish gray. The female resembles the male but her throat is dull cinnamon and the top of her head is duller. In both sexes, the bill, of moderate length and slightly downcurved, is dull black.

The species, of which several races have been recognized, is distributed through the high mountains from central Mexico to Honduras. In the former country it is found chiefly between 6,000 and 11,500 feet but has been known to descend as low as 1,000 feet (Friedmann *et al.*, 1950:176). In Guatemala I have encountered it from 7,000 to 10,000 feet, and I am aware of no record outside this vertical range.

My acquaintance with the Amethyst-throat is practically limited to the Sierra de Tecpam, where it occurred from the open, cultivated plains at the foot of the range upward through the zone of pines, oaks, and other broad-leafed trees to the cypress forest and cloud forest near the summit. On the plateau around 7,000 feet, I saw an Amethyst-throat beside a stream in the middle of the wet season, but I would hardly expect to find the species here in the drier part of the year. It became increasingly abundant as one ascended the Sierra, and its true home was in the damp mossy woods near the summit—not only the almost pure stands of cypress but, even more, the broadleafed cloud forest, where every gnarled trunk and tortuous branch and twig was sheathed in green moss, which also thickly carpeted the ground. In the more open parts of these cool, wet forests grew a tall shrubby salvia (*Salvia nervata*), whose flowers were the Amethyst-throat's chief source of nectar in its breeding season. Here, if I waited long and patiently in a favorable position, I would sometimes be rewarded with a flash of brilliant magenta reflected from the throat of a male who faced me squarely as he probed the long, furry, crimson corolla tubes. But only on the rarest occasions did his gorget appear as other than black.

Sometimes the Amethyst-throats ventured out to the more open mountain slopes to sip nectar from the red flowers of *Salvia cinnabarina,* and they also visited the tubular red blossoms of *Centropogon affinis.* In the wetter part of the year when the red salvias did not bloom, the Amethyst-throats depended largely on another red-flowered mint (*Satureja* sp.) and on the small, dull red flowers of the scandent shrub *Fuchsia minutiflora.* Although these hummingbirds seemed to be most attracted to red corollas, among the oak woods in the wet season they visited the large blue flowers of *Salvia cacaliaefolia.*

Voice and Courtship

In the mossy cloud forest, during the latter months of the year, each male Amethyst-throat rested on his habitual perch, from one to 10 feet above the moss-covered ground, and monotonously repeated a single squeaky note over and over again. The identical mournful note was reiterated, from 75 to 85 times per minute, as mechanically as the ticking of a clock, with never an attempt to introduce variety of pitch or phrasing into the utterance. One afternoon I watched an Amethyst-throat, perching only a foot above the ground in a dark ravine, deliver his single note 625 times without a pause. Even while flying, these hummingbirds sometimes sounded these same squeaky notes. They did not sing so continuously through the day as their neighbors in the more open places, the Green Violet-ears, but they were most vocal in the late afternoon and in cloudy weather. At such times the plaintive *squeak squeak squeak* arose on all sides in the still forest.

To call this tuneless squeaking a "song" may seem an abuse of the word. However, one may trace a complete gradation from the Amethyst-throat's mechanical calling through the Green Violet-ear's sprightlier utterance to the more varied and tuneful performances of such hummingbirds as the Wine-throated and the Band-tailed Barb-throat, which, if they only had a little more volume, we should include among the finest of avian music. Since the function of all these diverse utterances is evidently the same, if we call that of the Wine-throat a "song" we must, to be consistent, give the same designation to the whole series.

From time to time, the Amethyst-throat's song was suspended while he lazily stretched his wings and spread his tail in a brief moment of respite. At the conclusion of each long series of squeaks, he flew into the more open spaces of the woodland to draw the nectar from the crimson salvia blossoms. As he darted here and there among the shrubs, he sometimes made a noise which can best be imitated by holding a deck of cards against a table with one end slightly raised, then letting the cards slip rapidly downward over the thumb with a dry, buzzing sound. Occasionally one of the hummingbirds pursued another with a whirr of wings and a rapid metallic clicking, produced in a manner obscure to me. I could never decide whether these breath-taking chases were nuptial flights or the pursuit of a rival.

The Amethyst-throated Hummingbird's season for singing seemed to be regulated by the blossoming of the beautiful salvia whose nectar it preferred. On a short visit to the Sierra de Tecpam at the beginning of August, 1932, I found this shrub flowering rather freely and the Amethyst-throat singing much, in the middle of the rainy season. But in August of the following year, *Salvia nervata* bore no flowers and these hummingbirds rarely uttered a squeak. Toward the end of September the earliest blossoms were opening, and at the same time the Amethyst-throats sounded their monotonous notes more freely. By the middle of October, when the salvia flowers had become fairly numerous, the hummingbirds squeaked much of the day. They continued to sing through the early months of the dry season; but in February, when the salvias went to seed, they fell silent again. Unlike the Green Violet-ears, the Amethyst-throated Hummingbirds seemed equally abundant on the Sierra de Tecpam throughout the year.

Nesting

In late November of 1930, I found two nests of the Amethyst-throat, and two more in December of 1933. The sites of these nests were quite different. One was among the low, gnarled trees at the very summit of the mountain, 10,000 feet above sea level. The second, a few hundred feet lower, was in the cypress forest, on a pendent branch of a scrambling fuchsia that was draped with festoons of moss. The third was on the lowest limb of a cypress sapling that grew on a low bank beside a little-used logging road on "The Hummingbirds' Hillside"; it was in such a site as a Green Violet-ear might have chosen. The fourth nest was still lower on the Sierra, in the zone of mixed broad-leafed trees and pine, where in the midst of heavy, wet forest it was attached to a dangling spray of bamboo. In height, these four nests ranged from 4 to 6 feet above the ground, the highest being that on the bank above the logging road.

All the nests were substantial open cups, with thick walls composed largely or wholly of green moss, bound together by cobweb into a fabric of spongy consistency. The nest on the exposed mountaintop had the thickest walls, which seemed appropriate in this high, windy situation. Three of the nests were sparingly decorated with gray lichens on the outside, but the fourth lacked these encrustations. This last nest had, as inner lining, only two downy feathers from some larger bird. Another nest contained more feathers, and that amid the oaks and pines was abundantly lined with soft vegetable hairs of a rich chestnut-brown color and also a few fine mosses. Still another nest had no lining other than the moss of which its walls were made. These nests measured from 2 to 2½ inches in over-all diameter by 1¼ to 2¼ inches in height. The cavity was from 1⅛ to 1¼ inches in diameter and ⅞ to 1 inch deep. In the cloud forest where everything from the treetops to the ground was covered and draped with green moss, these green, mossy nests were not likely to be noticed unless the hummingbird flew from them as you walked by.

The two nests discovered in late November contained two eggs each, and those found in December held two nestlings. The two sets of eggs, white and of the usual narrow, elongate form, measured 14.3 by 9.1 and 13.9 by 9.1; 14.3 by 9.1 and 13.9 by 9.5 mm. The latest of the four nests contained two newly hatched nestlings when found on 27 December 1933. At their nests the Amethyst-throats were exceedingly shy, difficult to watch and to photograph.

PURPLE-THROATED MOUNTAIN-GEM

Lampornis calolaema

This beautiful highland hummingbird is of medium size, nearly four inches long, with a straight black bill of moderate length. The sexes differ greatly in coloration. The male's forehead and crown are glittering metallic emerald green; the rest of his dorsal plumage is metallic bronze-green, becoming more bluish green on the upper tail coverts. His deeply indented tail is dull blue-black. The sides of his head are dusky bronze-green, relieved by a conspicuous white streak behind each eye. His chin and throat are metallic violet or purple. The bright metallic green of the sides of his neck and his breast fades to brownish gray on the abdomen. On the female, the upper parts are green. The central feathers of her double-rounded tail are green; the other rectrices are green basally and blackish toward the end, with pale gray tips on the two or three outer pairs. The sides of her neck are blackish, with the auricular region bordered above by a conspicuous white streak that extends from behind the eye down the sides of the neck. The under parts of her body, from chin to abdomen, are cinnamon or tawny-ochraceous.

The Purple-throated Mountain-gem inhabits the highlands from western Nicaragua through Costa Rica to western Panama. According to Carriker (1910:542) it "is found throughout the highlands of central and northern Costa Rica, from about 3,000 feet upwards." I have found it abundant between about 5,400 and 8,000 feet, but have not happened to encounter it beyond this vertical range.

Food

Sometimes I have met the mountain-gem within the heavy, epiphyte-burdened forest; but more often it is seen in adjoining clearings, above all in pastures where scattered shade trees and stumps are overgrown with shrubs that at certain seasons bloom profusely. Especially attractive to the hummingbirds are the various epiphytic heaths, species of *Satyria, Macleania, Cavendishia,* and others of the subfamily Thibaudiae of the family Ericaceae. The tubular corollas of these colorful shrubs are mostly red or pink, often with white tips; and in *Cavendishia* they are borne among massed pink or red bracts which attract the eye from afar. These flowers have narrow mouths that are directed downward, so that to reach their nectar the hummingbirds must hover beneath them, with body vertical and bill pointed almost straight upward. As they hover so, the pollen sifts upon their heads from minute pores in the apices of the downwardly directed anthers; and the hummingbirds carry it from flower to flower as they make the rounds of their favorite plants. To hover with the axis of the body vertical apparently tires the hummingbirds, and often they cling to a convenient support while sipping the nectar. In June and July, the Purple-throated Mountain-gem was by far the most abundant hummingbird in the mountain pastures at the western end of the Barba massif, from 6,500 to 7,500 feet above sea level. Adult males were not rare, but birds in female plumage were far more numerous.

Bathing

In a deep, forested ravine on the "La Giralda" dairy farm in this locality, a low concrete dam had been built across a rivulet to form a small reservoir. Above the dam, the dark, still pool stretched, rapidly narrowing, back amid the overhanging vegetation which it reflected, until at the inlet it was completely overarched by the luxuriant shrubbery. Below the dam, the overflowing water glided down a short rocky incline, then plunged over the edge of a natural cliff, to fall vertically about 25 feet into a narrow chasm filled with lush vegetation. The opening around the head of the waterfall received far more light than reached the interior of the surrounding forest; and here, where in the middle of a clear day they enjoyed a few hours of sunshine, grew a variety of colorful shrubs and herbs: begonias with a profusion of pink blossoms, yellow calceolarias, lavender eupatoriums, a shrubby *Monnina* with deep blue flowers, *Dicliptera iopus* with pinkish magenta blossoms, a *Centropogon* with bright red corollas, and the elegant rubiaceous shrub *Deppea grandiflora* with clusters of golden flowers.

While I stood on the dam early on a morning at the end of March, long before the rays of the rising sun found their way into the deep ravine, a male Magnificent Hummingbird flew up and clung to the mossy rock at the head of the waterfall, where he pressed his head down into the shallow, smoothly flowing water and rapidly vibrated his spread wings. He did this over and over, wetting his glittering feathers in the cold, clear water. As I watched him below the level of my feet, his crown sent violet light to my eyes, while his throat was glittering green. Soon after he flew off into the forest, a male Purple-throated Mountain-gem came to bathe on the rim of the low central part of the dam, where the water overflowed. After his departure, a female mountain-gem, with her green forehead liberally dusted with pale pollen, bathed in the same place and manner.

The next bather was a Green Violet-ear, who performed his (or her) ablutions at the brink of the waterfall. Then a male mountain-gem clung to the nearly vertical face of the dam, doubtless using the verdant covering of moss for a foothold, and bathed in the falling stream. Next, several tiny Scintillant Hummingbirds of both sexes, but chiefly females or young birds, came to bathe. Some of them alighted on the top of the dam, where with spread tail they pressed themselves down into the shallow overflow, immersing themselves more deeply than the larger hummingbirds had done. For nearly an hour, there was a constant succession of bathing hummingbirds of these four species, but chiefly of the mountain-gems and the diminutive *Selasphorus.* Frequently the bathers were so close to my feet that I might have bent over and touched them. After eight o'clock attendance waned. Of the five kinds that I had so far found in this locality, only the big Violet Sabrewing had failed to appear.

A few mornings later I counted the bathers. From 6:15 to 7:15, 14 baths were taken, and likewise 14 in the following hour. Perhaps some of the hummingbirds bathed more than once. By species, the number of baths taken in the two hours was as follows:

Purple-throated Mountain-gem,	males 4, females or young 2.
Magnificent Hummingbird	males 11, females or young 1.
Scintillant Hummingbird	males 1, females or young 6.
Green Violet-ear	(sexes alike) 3.

When the bathing hummingbirds pressed their heads down into the flowing water, they often turned the head sideways, thus more effectively wetting the whole cheek. After one or several immersions of the foreparts, the bather beat its extended, raised wings into a haze, often making a humming sound, which was particularly loud in the case of the big Magnificent Hummingbird. Each hummingbird preferred to bathe alone, and sometimes one chased another away. But once a Violet-ear and a Magnificent bathed simultaneously, several feet apart. The baths of all these hummingbirds were very different from those taken in still pools by hummingbirds who fly down, partly immerse themselves, then without pausing fly up to shake themselves and preen on an overhanging branch. Aside from the four kinds of hummingbirds, the only bird that came to bathe at the dam was a solitary Mountain Elaenia. But this little gray flycatcher was as shy in my presence as the brilliant hummingbirds were bold; it took fright and departed when it had scarcely wetted its feathers.

When these observations were made in March, the dry season here on the Pacific slope of the Cordillera Central was at its height. My last visit to the dam was in early July, when the mountain forests were dripping after two months of rain. Now, from 7:00 to 8:00 on a sunny morning, only three baths were taken, one by a male mountain-gem and two by a female, probably the same individual on both occasions. No other species arrived. Now the birds had a far greater choice of bathing places, in the wet foliage and dripping moss no less than in the numerous rainy season rivulets.

Nest and Eggs

At Vara Blanca, 5,400 to 5,700 feet above sea level, I found six nests of the mountain-gem, the earliest of which must have been started at the end of December or beginning of January, as the young, well feathered when I first saw them, left on 15 February. In another nest, eggs were laid late in January, and in three nests they were laid in February. At the season when the hummingbirds incubated their eggs and fed their young, there were frequent, long-continued storms, when for days together a strong northerly wind drove the cold rain against this mountain slope, and sunny days were rare. At 7,500 feet, just across the continental divide from Vara Blanca, I found, on 26 March 1963, a nest with week-old nestlings, which had evidently hatched from eggs laid early in the same month. In January and February, scarcely any birds were breeding in this region, except hummingbirds and the Slaty Flower-piercer, which like hummingbirds subsists largely on nectar. In April and May, the height of the breeding season for the avifauna as a whole, we found only one nest of the mountain-gem, which was begun in early April and abandoned before completion. However, some mountain-gems breed later, for at the end of July I watched a female feed a full-grown fledgling, who might have hatched in June.

The six completed nests that I have seen ranged in height from 4 to 12 feet above the ground, with an average of 7 feet. That at the greatest altitude above sea level was in a narrow open space in a forest of oaks and other broad-leafed trees, with an undergrowth of tall, cane-like bamboos. It was situated 6 feet up on a slender, drooping spray of bamboo, where it was supported by a whorl of eight leafy branchlets no thicker than a matchstick, some of which were built into the nest's wall. The nest that was never finished was in a similar site on a bamboo at the edge of forest. The other nests were in clearings. Two were attached near the ends of stems of tall *gigante* grass than leaned over a bank beside a path which ran between a pasture and forest. These nests were well screened above by the same coarse grass that supported them. Another nest was in an elder tree in a pasture, and still another in a small lemon tree among coffee bushes behind a cottage.

These nests were substantial, thick-walled open cups, with incurved walls which diminished the chances that the eggs would be shaken out in the frequent windstorms. One nest was almost spherical in shape, with a flat top where the opening was situated. This nest was composed chiefly of buffy and tawny down, covered on the outside with green moss and a few gray-green foliaceous lichens. Strands of moss draped well below the bottom. Another nest was made chiefly of fine leafy liverworts, dead and brown, and large brown scales from the fronds of ferns, which predominated in the lining. The exterior was decorated with many long pieces of bright green, living hepatics, which hung below it, and a few fragments of lichens. A neighboring nest was of similar construction but had more gray lichens on the outside. One nest measured 2½ inches in diameter by 2 inches high.

Each of four nests contained, when found, two, tiny, white, strongly elongate eggs. Two others held each two nestlings. While one of the mountain-gems was incubating in her nest on the *gigante* grass at the pasture's edge, a female Slaty Flower-piercer built a nest amid the grass, 16 feet away. The hummingbird objected to the presence of the little honeycreeper, whose body was scarcely larger than her own, and her bill far shorter. Could she have instinctively recognized as a rival a bird who subsisted largely on nectar, just as she did, although it obtained the sweet fluid by a method that few hummingbirds follow? During her breaks in incubation, the mountain-gem darted repeatedly at the building flower-piercer, almost striking her, if not actually doing so. But undismayed by these swift onslaughts, the builder went about her business, paying scarcely any attention to her belligerent neighbor. While at her nest amid the close-set grass stems, the flower-piercer was secure from attack, for the long-winged hummingbird could not reach her. Once when the flower-piercer was in the open pathway below the nest of the hummingbird, the latter darted directly from her eggs to assail her.

At one nest, an egg hatched 15 days after I found the completed set.

Nestlings

One female mountain-gem, who while incubating would permit me to approach within arm's length, became even bolder on the day her nestling hatched and did not fly away until my advancing hand was within 3 inches of her. At the nests I found in 1938, I saw no indication that a male was interested; but I made no long-continued watch. In later years, after evidence was published by Moore (1947) and Schäfer (1954) that some male hummingbirds depart from the procedure usual in the family and help to attend nests, it became of interest to investigate this point for as many species as possible. Accordingly, when I found the mountain-gem's nest in 1963, I spent a morning watching it. From 6:10 to 11:10 a.m. on 29 March, the female came 12 times to feed the two nestlings. As a rule, on each visit she fed one nestling twice and the other nestling once. Each feeding or uninterrupted act of regurgitation lasted from about 3 to 8 seconds, usually 4 or 5 seconds. Although the morning was clear, here in the shade of the oak forest at 7,500 feet the air was uncomfortably chilly; yet the nestlings, still in pinfeathers without expanded plumage, were not brooded for a single minute. Only their mother attended them; no male was ever seen near this nest. All morning she uttered no note audible where I sat unconcealed 20 feet from the nest, my presence seemingly ignored by her.

At one nest, the nestlings were left exposed during the night when they were respectively 12 and 13 days old. At lower altitudes, young Scaly-breasted Hummingbirds may sleep without the maternal coverlet when only 8 to 11 days old and still almost naked; but at greater heights where nights are frosty, White-eared Hummingbirds are brooded until they are 17 or 18 days old and clothed with feathers. From one nest of the mountain-gem, two young flew when respectively 22 and 23 days old. In a neighboring nest, where only one egg hatched, the lone nestling departed at the age of 23 days. At this age the sexes can already be distinguished, for males have green on their breasts and females are buffy.

While I watched a Quetzal's nest at the end of July, a young male mountain-gem rested near me while his mother came at intervals to regurgitate food to him. He already resembled the adult males, although his plumage was less brilliant: the metallic blue-green of his forehead was less intense than on adults, and the patch of metallic violet on his throat was much smaller. At times this young hummingbird would tilt up his head, swell out his throat, and sing with closed bill. His notes were rather sharp, but so weak that I could scarcely hear them at a distance of three yards. Despite the slightness of this performance, it seemed to give the youngster considerable pleasure, for he repeated it persistently. Years later, I heard a young Scaly-breasted Hummingbird sing in similar fashion while still receiving nourishment from his mother, as he did until he was at least 65 days old and had been out of the nest 41 days. Adult male Scaly-breasts sing persistently in courtship assemblies, and the foregoing observation suggests that mountain-gems may do so, too; but if so, I have never heard them, nor learned anything about their courtship.

Family RAMPHASTIDAE

BLUE-THROATED TOUCANET

Aulacorhynchus caeruleogularis

This small toucan, about 11 inches in length, is largely green, with a patch of dark blue covering the lower cheeks, chin, and throat. The under tail coverts and the ends of the tail feathers are deep cinnamon-rufous or chestnut. The relatively huge bill is variegated with four colors. The greater part of the upper mandible is yellow faintly tinged with green, but its base and cutting edge are black, as is the whole lower mandible. There is a small patch of dull red at the base of the culmen, and a broad white line outlines the bill at its base, where it joins the head. The sexes are alike in coloration, but males average larger, and in some pairs the male's bill is noticeably longer than the female's.

The Blue-throated Toucanet is confined to the highlands of Costa Rica and Panama. In the former republic, I have found its nests from 3,700 to 7,000 feet above sea level; the absence of records from greater altitudes probably reflects only the small amount of ornithological work done in the higher parts of the country. In the coolest months the toucanet wanders lower: I occasionally met it at about 3,000 feet on the slopes above the Río Buena Vista on the northern side of the basin of El General, but only once in many years have I seen it as low as our farm at El Quizarrá. This was on 28 November 1963, when I found a lone toucanet at about 2,400 feet. On the Caribbean slope the toucanet may descend even lower, down to 1,500 feet (Slud, 1964: 183). While I resided at Montaña Azul, I noticed that the toucanets, which were quite abundant around 5,500 feet in July, became rare in September and October, then gradually increased in numbers during the early months of the following year. Evidently the local population had descended the mountain slopes to pass the stormiest season in warmer regions. Below 4,500 or 5,000 feet, the toucanet mingles with the big toucans of the genus *Ramphastos* and the middle-sized araçari toucans of the genus *Pteroglossus*; above this altitude it is the only representative of its family in Costa Rica.

Like other toucans, the Blue-throated Toucanets associate, when not breeding, in small flocks, rarely consisting of more than six or eight individuals, which straggle along from tree to tree one behind the other, instead of moving as a compact group. They seem most at home high in the trees of the wet, epiphyte-laden mountain forests, but they often come forth into adjoining shady plantations and pastures with scattered trees, to forage and even to nest, if suitable sites are available. Restless and excitable, they scold a human intruder in tones which often resemble the chatter of an angry squirrel, at the same time holding their tails and their great-billed heads in various angular, ungraceful attitudes. Throaty, frog-like croakings, dull barks, and dry rattles are their most common utterances, but they sometimes give voice to softer notes while nesting. They eat a variety of small fruits by the toucan's usual method, plucking them one by one in the tips of their great mandibles, then tossing them back into their throat with an upward jerk of the head. They also consume a variety of insects and, in the breeding season, they vary their diet with the eggs and nestlings of other birds.

My earlier account of the Blue-throated Toucanet was based upon six nests that I found at Montaña Azul in 1938 (Skutch, 1944b). Since this was written, I have found five more nests and gathered some new information about the bird's breeding habits, especially the role of the sexes in nest building and incubation. The following account is limited to observations not previously published.

Nest Building

According to Ridgway (1914: 358), the lengths of the bills (culmen) of adult males range from 62.5 to 73 mm. and average 66.8 mm., while those of females range from 52 to 61 mm. and average 57.2 mm. From these measurements it appears that the differences in the sizes of the bills of a particular pair of toucanets would not always be sufficiently pronounced to be recognized in the field, but whenever one member of a pair has a bill noticeably longer than the other this would be the male. At the nests which I studied carefully in 1938, I failed to distinguish the sexes, possibly because I did not then know what difference to look for. But in a pair of toucanets which I watched at Cañas Gordas in 1964, one member had a much longer bill than the other. This pair, especially the small-billed partner, worked hard at enlarging nest holes in decaying trunks. At Montaña Azul one pair of toucanets evidently widened the doorway of a hole which they wrested from a pair of Hairy Woodpeckers that had carved it; and another pair of toucanets, as I concluded from indirect evidence, deepened a chamber in which a Golden-olive Woodpecker had previously slept. But in neither case did I actually watch the toucanets at work.

The pair of toucanets at Cañas Gordas first tried to carve a hole in a vertical band of decaying wood in the side of the trunk of a tall, slender, living tree standing in a pasture not far from forest. The point where they worked was 15 feet above the ground. Here an irregular, roundish opening, too narrow to admit a toucanet, gave access to a cavity which penetrated rather deeply in a horizontal direction but seemed not to descend below the doorway. The excavation appeared to be newly begun, apparently by the toucanets themselves. From 9:19 to 10:49 a. m. on 23 March I watched these birds enlarging the hole. The female took 11 spells at the work, ranging from 2 to

FIG. 1. Shady pasture at Finca "Loma Linda," near Cañas Gordas, Costa Rica, 3,800 feet, where Blue-throated Toucanets nested. The largest tree is a Mexican elm (*Chaetoptelea mexicana*) about 170 feet high; the palms are *Euterpe* sp. May, 1964.

10 minutes and totalling 56 minutes. The male took only three spells, lasting 3, 5 and 5 minutes, and even while at the hole he rarely exerted himself as much as his companion, whose shorter bill seemed a more effective tool for this work. While carving, she clung with her feet clasping the trunk below the doorway and her tail pressed against the trunk for much of its length. She pecked or hammered much at the wood, so hard that I could hear the sound of the blows 100 feet away. She often, perhaps always, delivered the blows with her mandibles slightly parted. Instead of throwing her whole body into the stroke, as woodpeckers do, she moved only her foreparts when she struck the wood. She also seemed to bite away the rotting wood with her bill, but this was difficult to see, for her head was inside the hole. The loosened particles dropped out on her breast, from which they slipped off to the ground. While one partner worked, the other rested much of the time in a tall shrub of *Ardisia* that grew in front of their hole, from time to time eating the juicy purple berries.

After 10:49 I saw no more work done in the morning, but in the afternoon of the same day I watched the female work for three spells lasting 5, 3, and 3 minutes, while her mate looked on. The following morning I watched for the birds to resume their task without seeing them, probably because of the strong wind which blew all day. A few days later it was evident that the toucanets' undertaking had been abandoned unfinished, apparently because the wood proved to be too hard for them to work.

On 29 March, I learned that a toucanet had slept the preceding night in a hole about 30 feet up in a massive, decaying trunk in the pasture, a few hundred feet from the excavation which was now abandoned. The doorway of this hole, which looked old, was partly screened by the great, glossy, perforated leaves of *Monstera,* an epiphytic aroid. Later that same morning, I watched the female toucanet remove five overflowing billfuls of wood particles from this cavity, carrying each load to a neighboring tree before she shook it from her bill. It interested me greatly to see a toucanet remove the excavated material to a distance, just as hole-carving barbets do (Skutch, 1944a), instead of simply throwing the wood particles through the doorway, as, in my experience, woodpeckers invariably do. While the female toucanet worked, her mate rested on a petiole of the *Monstera* below the doorway, sometimes preening and gaping. Then he clung to the doorway and looked around, but did not enter. Later in the morning he did go inside and stayed for four minutes, during which the female twice entered and left the hole and once clung to the doorway without going in. When finally he emerged, I detected nothing in his bill.

The dozen nests of the Blue-throated Toucanet that I have seen over the years ranged in height from 7 to an estimated 90 feet above the ground. All were in holes in dead or dying trees, most of which stood in clearings a short flight from the forest, although a few were just within the edge of the woods. My failure to find a nest deep in the forest probably reflects merely the greater difficulty of discovering nests high in close stands of trees. One toucanet's nest cavity was 18½ inches deep, measured from the lower edge of the doorway. The entrance to the hole is oval in form with the long axis horizontal; the width varies from 2⅛ to 2½ inches and the height from 1¾ to 2 inches. The bottom of the cavity is always devoid of a soft lining. The eggs are pure white, three or four in a set.

Incubation

On the evening of 1 April, I found the male resting in a tree near the trunk on which the *Monstera* grew in front of the nest hole, from time to time stretching his wings. After 20 minutes his mate arrived, and he accompanied her to the aroid in front of the nest. She entered to stay for the night; he flew away. Soon afterward, three Fiery-billed Araçaris entered a hole higher in the same trunk, where for some nights they had been lodging. On succeeding evenings, events occurred in the same sequence. The female toucanet did not even look from her doorway when the larger toucans entered their dormitory some yards above her. Each evening the male toucanet flew off through the pasture after seeing his mate installed in the nest chamber, but

I could not discover where he passed the night. Although araçaris sleep alone only in the hole in which incubation is in progress and at all other times lodge in pairs or in larger groups, in the nest cavity or some other hole (Skutch, 1958b), I have never known more than one toucanet past the nestling stage to occupy a hole. Apparently, when not incubating or preparing to incubate, or else brooding nestlings, toucanets roost in the open rather than in holes.

One evening, while the pair of toucanets rested near the hole in which they were preparing to nest, the large-billed bird tried to mount the one with the smaller bill; then the latter entered the cavity for the night. Next morning, near the nest, the large-billed member of the pair gave the other a lauraceous fruit, which she ate. Then he billed her neck and tried to mount her, but she resisted. Thus I had confirmation from behavior that my identification of the sexes by bill size was correct. I have also seen a Fiery-billed Araçari and a Chestnut-mandibled Toucan feed a companion, evidently its mate; but in these cases I could not tell which was the male and which the female. After the feeding, one of the toucanets scratched its head by raising its foot outside its wing, which it held folded against its body, just as woodpeckers do. In piciform birds I have not noticed the over-the-wing (or inside-the-wing) head scratching so prevalent in passeriform birds and hummingbirds.

Incubation evidently began in the hole screened by the aroid about 4 April, but the attendance of the parents was at first so desultory that I was not convinced that they had eggs until some days later. I watched the toucanets incubate through the afternoon of 15 April and the mornings of 16 and 18 April, but the record made in the afternoon is imperfect because I failed to notice all the birds' movements. The incubation pattern of toucanets is more complicated than that of birds which sit more patiently, and for clarity I have summarized the records of the two mornings in Table 1. When the record for each morning is read across and downward, as one reads a book, it gives the actual sequence of sessions by the two parents and the intervals when both were absent, all in minutes. On the morning of 16 April, the female, who had occupied the nest through the night, sat until her mate came at 5:39, long before sunrise, to replace her. He then sat for only 10 minutes. On 18 April he arrived at 5:40, but instead of staying to incubate after the female left, he followed her from the nest tree and did not return to attend the eggs until seven minutes later. Then on both mornings the two partners alternated on the nest several times, with intervening periods of neglect.

After the male replaced the female at 8:10 on 16 April, he alone was in charge of the nest for the next four hours and 12 minutes. But far from incubating continuously, as many a smaller bird would have done, he broke this long period into eight sessions on the eggs, with intervening recesses. The first session lasted 53 minutes, but the others

TABLE 1
INCUBATION BY A PAIR OF BLUE-THROATED TOUCANETS

16 April, 1964. 5:39 a.m.-12:22 p.m. Intervals in minutes			18 April, 1964. 5:40 a.m.-12:19 p.m. Intervals in minutes		
Male	Female	Neither	Male	Female	Neither
10		5			7
	21	5	5		7
1		4		37	
48	25	4	11	30	
	28		38		9
53		4	27		1
21		7	29		2
27		10	21		9
25		22		18	4
20		12		14	2
21		3		13	9
6		8		28	10
13		1		19	9
	began			19	21
			began		
245	74	85	131	178	90

Average of 17 sessions by male, 22.1 minutes; range 1-53 minutes.
Average of 11 sessions by female, 22.9 minutes; range 13-37 minutes.
Average of 24 intervals of neglect, 7.3 minutes; range 1-22 minutes.

were much shorter. The decreasing length of his sessions revealed his increasing restlessness as his mate's absence was prolonged. Sometimes he passed his whole recess resting on a petiole of the *Monstera* that grew in front of the doorway, often stretching and preening. At other times he perched in a neighboring tree, whence he could see the nest, but occasionally he flew farther off. His longest period off the eggs lasted 22 minutes, for only 11 of which was he out of sight of the nest. Although too restless to sit for even an hour at a stretch, he was obviously ill at ease away from the nest, unless he knew that his mate was there.

On 18 April, the toucanets divided the task of incubation more equally. From 7:17 to 9:32 the male was in charge, taking four sessions separated by short recesses. Then at 9:33 the female took over and sat for six intervals, broken by five rests lasting from 2 to 10 minutes. At 12:10 p. m. the male returned and at 12:19 he entered the nest. When not in the nest, the female also spent much time resting in front, guarding it, but sometimes she was out of sight during the whole of a brief recess. Taking the two mornings together, in nearly 13½ hours of watching the male incubated a total of 376 minutes, the female 252 minutes, and the eggs were neglected for 175 minutes. The two parents together kept their eggs covered for only 78 per cent of the time, which is no better than many a smaller bird does alone. Nevertheless, they were slightly more attentive than another pair of Blue-throated Toucanets that I watched years earlier, and far more constant than a pair of the considerably larger Fiery-billed Araçaris (Skutch, 1944b:139-141; 1958b:213-214). Although I had long known that, for such large birds, toucans of various species incubate most restlessly, until I studied this nest at which I could distinguish the sexes, I was not aware that the same parent may take a number of consecutive sessions (up to eight at this nest) before the other parent relieves it. In other kinds of birds in which the sexes share incubation, the sessions of the two tend to alternate.

A question of long standing that was answered at this nest was: Which sex among toucans takes charge of the eggs by night? I would not have been surprised if, as in the related woodpeckers, the male had incubated through the night; but repeated observations proved that the female toucanet did so. The order Piciformes is far from uniform in this matter. We now know that in at least one jacamar, the Rufous-tailed, and one toucan, the Blue-throated Toucanet, the female alone occupies the nest by night while it contains eggs and young. In at least one puffbird, the White-whiskered Soft-wing, the male occupies the nest, which as far as known is universally true of woodpeckers, except in those genera *(Tripsurus, Picumnus)* in which the mated pair sleep together in the hole at all times. In the Prong-billed Barbet, the male and the female also sleep together in the nest while it contains eggs and young.

While the Blue-throated Toucanets were incubating in the daytime, I sometimes heard tapping or hammering coming from their trunk. The female, after she entered the hole for the night, tapped much while daylight faded. From the character of the sound, I suspected that the toucanets were working to enlarge their nest chamber. However, considering their evident distaste for the task of incubation, they may have been tapping on their wall merely to relieve their boredom, as a man beats a tattoo with his fingers on the table. When one partner came to take a turn at incubation and found the other inside, I heard a rattling sound as the latter emerged, pushing past the newcomer. I could not tell whether one sex or both made this noise. In the afternoon while the sun shone hotly, the toucanet in charge of the eggs spent much time with its head in the doorway, instead of incubating.

Nestlings

At an accessible nest, I had found the incubation period of the Blue-throated Toucanet to be 16 days—the only determination of this period for a toucan of which I am aware. The newly hatched young, pink-skinned and absolutely naked, had tightly closed eyes and bills of which the lower mandible was both longer and broader than the upper, as in woodpeckers, jacamars, barbets, and kingfishers. As in other nestlings that grow up in unlined holes and burrows, each heel of the young toucanets bore a prominent callose pad, studded with papillate projections, which would save this joint from abrasion against the hard floor of their nursery.

At the nest behind the *Monstera* leaves, the toucanets were bringing the food for nestlings by 20 April, proving that the eggs had been laid considerably earlier than I had supposed. I passed the afternoon of this date watching the parents attend their newly hatched young, of undetermined number. From 1:05 until the female entered the hole for the night at 5:55, the male brooded for seven intervals, ranging from 2 to 23 minutes and totalling 56 minutes. The female brooded the nestlings 11 times, for periods ranging from 2 to 29 minutes and totalling 113 minutes. The nestlings were alone for 11 intervals ranging from 1 to 31 minutes and totalling 121 minutes.

On this afternoon, the female took food into the nest 10 times, the male 7 times. Sometimes the article held in the tips of the parent's mandibles was unrecognizable, but often it was clearly a small fruit and sometimes an insect. The female's contributions tended to be smaller and better mashed than the male's, which on two occasions were too big for the nestlings to swallow, so that he emerged from the nest still holding the food and then ate it himself. Another time he came with empty bill, perched in a tree near the nest, and regurgitated a large green fruit. He held it in his bill as though considering whether it would do for a nestling, then swallowed it again, flew away, and in a few minutes returned with a much smaller fruit, which he took inside the nest. This reminded me of an amusing episode that I had witnessed long before. A parent toucanet feeding older nestlings arrived with a very large lauraceous fruit in its bill and took it into the nest. Presently the parent tried to leave the hole, but got stuck halfway through the doorway; push and squirm as it would, it could not pass through. My suspicion of what had happened was confirmed when it regurgitated the big fruit, which it had evidently swallowed after it found that the nestlings could not. With its girth reduced by this expedient, the toucanet slipped through the doorway without difficulty, holding the fruit in its bill. I have seen toucanets' nests at which the parents could barely squeeze through the orifice even at the end of a spell of incubation.

That afternoon, in a lauraceous tree 50 yards from the nest behind the *Monstera,* I watched the parent toucanets croak at a Fiery-billed Araçari, probably one of the three who slept in the hole above the nest. Then the toucanets tried to drive the larger toucan away, but it turned the tables and chased them from branch to branch. These toucanets also worried a pair of Golden-naped Woodpeckers nesting high in a neighboring tree. On several occasions they chased the woodpeckers as the latter approached their own nest hole. The male toucanet repeatedly stuck his head into the woodpeckers' doorway, trying to reach the nestlings. Finding that he could not, he attempted to enlarge the doorway so that he could enter; but the wood around it was too solid for him to tear away.

The toucanets also worried the Chestnut-headed Oropéndolas in a neighboring small colony. One afternoon I watched a toucanet investigating the long, woven nests that hung in a high treetop. The toucanet clung to the sides of the swinging pouches, sometimes upright and sometimes with head downward, and once it entered a nest but evidently did not descend to the bottom. While the toucanet searched among the pouches, a pair of Piratic Flycatchers, preparing to nest in one of them, protested violently. Again and again they darted, with snapping mandibles, close by the bird so much larger than themselves, and once one of them knocked out some of its green feathers. But the oropéndolas had already finished breeding, and the would-be nest-robber, finding neither eggs or nestlings in the pouches, finally flew off, leaving the flycatchers calling vociferously.

At an earlier nest of the Blue-throated Toucanet, the young did not leave until they were at least 43 days old. Long before they reached this age, the nestlings vanished from the nest behind the aroid. After this occurred, I saw the male of this pair give two pieces of food to his mate. Evidently they were preparing to try again to rear a brood; but if they started another nest, I could not find it. As far as I know, lowland toucans raise only a single brood, and Fiery-billed Araçaris which had lost eggs or nestlings did not renest in the same year. But, as I learned at Montaña Azul, the more prolific toucanets may produce two broods in a season, which accounts for their abundance in the wet, epiphyte-laden mountain forests.

Family DENDROCOLAPTIDAE

OLIVACEOUS WOODCREEPER

Sittasomus griseicapillus

Although only about six inches long and small for its family, the Olivaceous Woodcreeper has the features typical of the Dendrocolaptidae: slender form; shades of brown predominating in the plumage; no sexual differences in coloration; slender bill for probing into crevices in the bark of trees; tail of 12 stiff rectrices, the shafts of which have sharp, projecting, downcurved tips, well fitted to engage the bark and support the bird as it climbs up trunks. The Olivaceous Woodcreeper is grayish olive on the pileum and hindneck, lighter grayish olive on the sides of the head and under parts, except the under tail coverts. The back and shoulders are russet, which brightens to deep cinnamon-rufous on the rump, upper tail coverts, and tail. The under tail coverts are also cinnamon-rufous, and there is much of this color on the remiges, which are variegated with areas of dull black and buff, visible chiefly when the wings are spread.

The Olivaceous Woodcreeper ranges from central Mexico to Bolivia, northern Argentina, and the island of Tobago. Over much of this vast territory it inhabits the tropical lowlands, and I have found it near sea level in northwestern Costa Rica, where the dry season is long and severe. But in the more humid parts of Costa Rica where I have seen most of this species, it prefers higher elevations. On the Pacific slope of the Cordillera de Talamanca, it rarely descends as low as 2,500 feet in El General, and it is far from common even at 3,000 feet. In the forests near the Panamanian border, it was common around 4,000 feet and even more abundant at 5,000 feet, above which I have not been able to trace it. On the Caribbean slope of Costa Rica, it lives chiefly between 2,000 and 5,000 feet; I rarely saw it as high as 5,500 feet at Vara Blanca. In Venezuela, where the species is represented by a number of races, it occurs from the lowlands up to nearly 7,000 feet (Phelps and Phelps, Jr., 1963:44-45).

This little woodcreeper has less character than some of its larger relatives. It seems to lead a monotonous life; one sees it nearly always industriously climbing up the trunks of forest trees, or sometimes outward along the branches, using its tail as a prop for its slender body, and with its slender bill plucking from the bark articles of food that are usually too small to be detected from the ground. Sometimes it darts outward into the air to capture some insect that has attempted to escape by flight. Its call is a very sharp, fine, rapid trill, resembling that of the Plain Xenops, but more attenuated. A solitary bird of the forest, it rarely joins the mixed flocks, and it associates little even with others of its own kind. Considering the wide range of the Olivaceous Woodcreeper, surprisingly little has been written about its habits; the collectors' notes on this species are as a rule even briefer than those on many others. For this reason, it seems proper to report here the little that I have been able to learn about it.

Nesting

As I walked along a forest trail near Cañas Gordas toward the middle of the morning of 25 March 1964, I saw an Olivaceous Woodcreeper carry three small dead leaves into a wide opening about 40 feet up in the side of a tall dead trunk of a palm (*Euterpe* sp.) Although I waited a good while, the bird brought no more leaves. During the succeeding two weeks, I looked in vain for activity at this site. On 7 April, however, I saw the woodcreeper carry another piece of material into the hollow trunk. Late on the morning of 12 April, it carried in two or three pieces. Whenever I saw the builder, it was alone. From time to time, it uttered its fine trill.

On 18 April, the woodcreeper emerged from the palm trunk when I tapped on its base, and the same thing happened on the following day. Evidently incubation had begun. Although some nesting woodcreepers prefer a cavity with an opening barely wide enough for them to squeeze through, the pear-shaped gap in the side of the palm trunk by which this woodcreeper entered and left was far higher and wider than it needed to be for so thin a bird. Above and below it were smaller gaps in the shell of the hollow trunk, which was far too weak to support a man's weight. Hence I could not learn how far below the doorway used by the woodcreeper its nest was situated. On the lower part of the trunk grew a scandent begonia (*Begonia glabra*) with pale green, glossy leaves and dry seed pods. This nest was at an altitude of about 3,600 feet.

I spent most of the morning of 21 April watching this nest, where the woodcreeper seemed to be incubating. As during the weeks when leaves were occasionally taken into the palm trunk, I at no time saw more than one *Sittasomus* in the vicinity. Apparently only the female was incubating. Unfortunately, I missed one departure from the nest, making my record imperfect; but I timed six sessions and as many recesses. The sessions lasted 2, 27, 9, 19, 58, and 33 minutes; the recesses, 7, 8, 11, 13, 13, and 25 minutes. Each time the woodcreeper left the nest, she emerged from the doorway and climbed several feet to the top of the trunk before she flew. On returning, she invariably alighted on the trunk several yards below the entrance and climbed up to it. I never saw her approach or leave the doorway by flying; even when (as on other days) she was disturbed by tapping on the base of the trunk, she climbed higher before she flew away. On every return to the nest but one, she brought a small dead leaf or a fragment of a leaf to add to the accumulation in the cavity. Sometimes she slipped through the high doorway in an upright position and backed down into the cavity tail-first until out of sight; but at other times she pivoted over the sill and seemed to descend head-first. On this morning, she invariably approached and left the trunk in silence.

A few days later, before I saw any food carried into the palm trunk, the woodcreeper failed to respond to my tapping on the base. Evidently some mishap had befallen the eggs; perhaps the rains which now began to fall damaged the nest, since the hollow trunk was open at the top and permitted their entry.

Some woodcreepers live in pairs throughout the year, whereas others are solitary even while they nest. At a number of nests of two species of *Lepidocolaptes,* the Streaked-headed and the Spotted-crowned Woodcreepers, I have seen that both sexes share all the work, from building to feeding the young; and at one nest of the Wedge-billed Woodcreeper, both sexes incubated. But I have consistently failed to find a second parent taking any interest in numerous nests of the Tawny-winged Dendrocincla. Evidently *Sittasomus,* like *Dedrocincla,* is a woodcreeper in which lasting pair bonds are not formed. This needs to be confirmed by observations at other nests; but it may be long before another nest is found by one able to study it.

Family FURNARIIDAE

SPOTTED BARBTAIL

Premnoplex brunnescens

As in many other species of ovenbirds, the Spotted Barbtail is clad in shades of brown and the sexes are indistinguishable. It is about five and a quarter inches long. The upper plumage is deep brown, more olive or grayish olive on the top of the head, darker on the tail, and with faint dusky bars on the back. There is a buffy superciliary streak above the dusky cheeks. The chin and throat are deep buff, and the remaining under parts are olive, nearly everywhere conspicuously spotted and streaked with buff. The shafts of the 12 stiff tail feathers project in slender barbed points. The short, slender, slightly curved bill has a black upper mandible and a yellowish lower mandible.

The Spotted Barbtail ranges from Costa Rica to Bolivia and Venezuela. In Costa Rica it occurs, according to Carriker (1910:648), from about 2,000 feet up to timber-line on the high volcanoes, but it is not common below 5,000 feet except on the Caribbean slope, where it is abundant down to about 3,500 feet. I have met the bird at points ranging from about 3,500 to 7,000 feet; but only around 5,000 feet on the Pacific slope of the Talamancan Cordillera in extreme southern Costa Rica did I find it common. In the Santa Marta region of Colombia, the species extends vertically from 5,000 to 9,000 feet but is rare above 7,000 feet (Todd and Carriker, 1922:292), and in Venezuela it occurs from about 3,300 to 8,200 feet (Phelps and Phelps, Jr., 1963: 62-63). In eastern Ecuador I found this barbtail not uncommon in the wet forests near Puyo at about 3,000 feet.

The Spotted Barbtail is confined to wet, mossy, epiphyte-encumbered mountain forests and is especially fond of their darkest and dampest parts, the deep ravines into which only stray beams of sunshine find their way. It travels through the lower levels of the woodland in mixed parties of small birds, usually only a single adult barbtail to a flock. It hunts chiefly on the trunks and branches of small trees and the lower parts of the trunks of great ones, never ascending into the sunny treetops. It appears to subsist wholly on insects and other small invertebrates, in the quest of which it is so incessantly active that it is most difficult to keep in view as it creeps over mossy boles and branches in all positions, head or tail, back or belly, or either side uppermost. Although this ovenbird sometimes works up vertical trunks with its body upright and its tail pressed against the bark, in the manner of a woodcreeper, on the whole it seems to depend little upon its strongly barbed tail feathers for support. Rarely it perches crosswise on a thin branch, in typical passerine fashion. Its usual note is a slight, thin *chip,* which may be rapidly repeated to form a dry rattle or trill.

Nesting

The very strangeness of some of the birds which, like the Spotted Barbtail, live obscurely in the dim depths of the dripping mountain forests challenges us to cultivate their acquaintance; but often they are so retiring, the range of their activities and vocalizations appears to be so narrow when compared with those of birds of sunnier habitats, they seem so deficient in character or "personality," that we feel poorly rewarded for our strenuous efforts to know them more intimately. The very profuseness of these wet forests, the wealth of opportunities which they offer for the support of bird life, appears to force many of the birds to become narrow specialists in order that the maximum number of kinds may thrive in the same area. But just as we conclude that these elusive mountain birds will hardly reward further study, we may have the rare good fortune to stumble on one of their nests, which is so excellently concealed, and of such unexpected form, that it amazes and delights us. We may suspect that the whole purpose of the bird's existence, its one claim to distinction, is the production of a nest that is altogether unique.

Near Agua Buena in extreme southern Costa Rica, at an altitude of about 3,700 feet, a massive, steeply inclined, rotting log, verdant with mosses and ferns and other epiphytic growths, served as a precarious footbridge over a shallow brook flowing through a deep wooded ravine between cow pastures. As the boy who was guiding me started to cross this log, a small bird flew from it and vanished without giving me a view adequate for identification. After much searching along the part of the log from which the unknown bird appeared to have taken flight, I found a green mossy nest perfectly concealed amid the green moss and ferns. It was simply a pocket that opened downward; it could not hold an egg, nor could I see how it could be converted into a structure in which a brood might be reared. I concluded that it was only a dormitory nest; but its distance from my residence, over a rough trail, prevented nocturnal visits for the investigation of this point.

When next I passed this way, a fortnight later, the mossy structure had been altered in an unexpected manner and held two eggs. The form of this nest, which I later removed as a coherent green mass, is best appreciated by reference to Figure 2. The bulky mass measured 8½ inches in height, 7½ inches from front to back, and 4½ inches from side to side. It was composed of mosses, liverworts, and fine, dark-colored rootlets, evidently of epiphytic plants. This compact structure was penetrated from the bottom by a vertical tube, 1½ inches in diameter and about 3¼ inches long. This tubular entranceway led to the chamber, where in a shallow concavity in the top of the main mass of material the eggs lay. This chamber was about 2¼ inches in di-

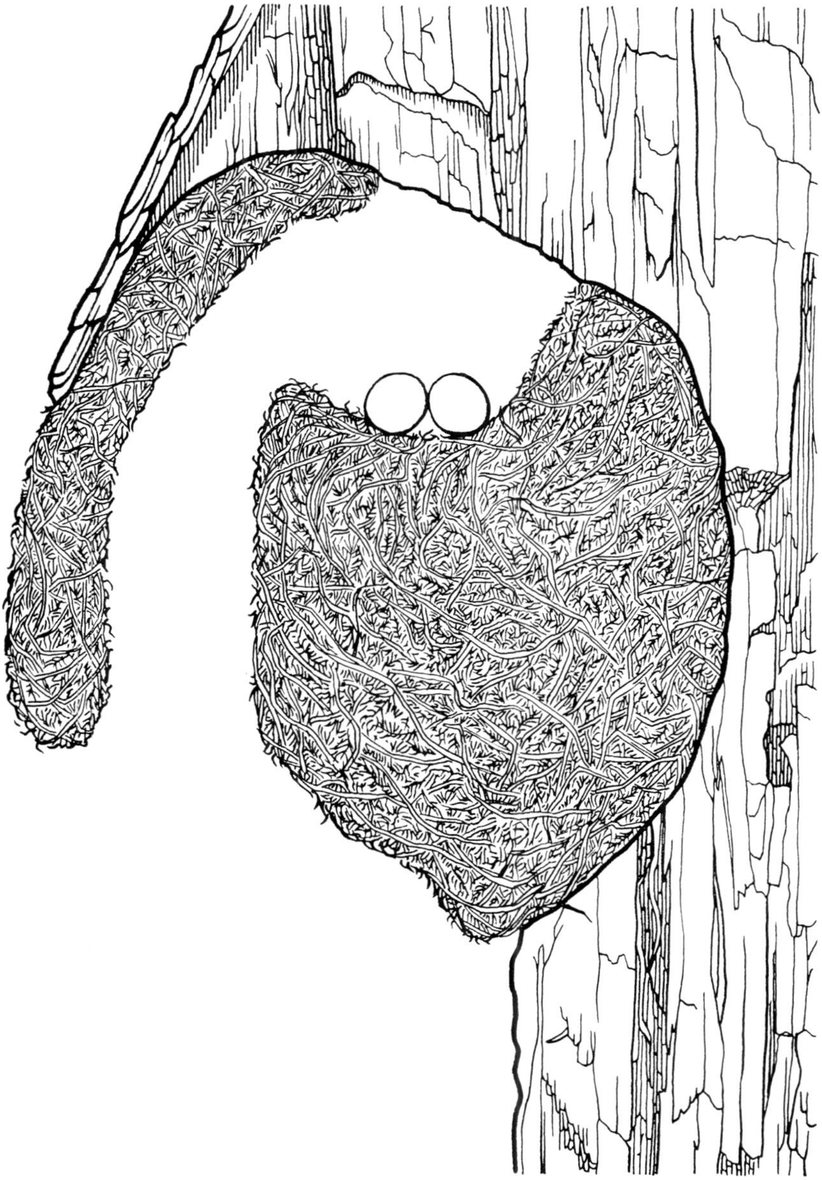

FIG. 2. Sagittal section through nest of Spotted Barbtail, *Premnoplex brunnescens*. One half natural size.

ameter by 3 inches high, and its ceiling was the rotting wood of the supporting log, with no covering. Accordingly, when the nest was removed from its niche, it was found to be an irregular mass of moss and other materials, penetrated from top to bottom by a hollow of the form shown in the figure.

The front of the niche in which the nest was set was partly covered by a flap of bark which kept the completed structure from falling forward. But I could not understand how, during construction, the nest had been held in place before the increasingly heavy mass of green, wet plant tissues was made to fill the niche snugly. Possibly the first materials brought by the birds were entangled in moss that grew on the walls of the niche. If so, this attached moss offered no perceptible resistance when I carefully removed the nest; but by this time it was dying as a result of being deeply embedded in the mass of nest material.

On 30 May 1964, this nest held two pure white eggs, which I could see lying in their shallow hollow only when I stood in the stream below and looked up the entrance tube, through which I pushed a tiny mirror and an electric bulb attached by a cord to a flashlight.

Kicking the end of the log that rested on the lower bank of the brook, and shaking it gently, did not make the incubating barbtail leave the eggs; but it would fly out if I waded into the stream and touched the nest itself or the log close beside it. The first time that I did this, the bird darted away so swiftly that I again failed to identify it.

On 31 May, I watched one of the barbtails enter the nest at 7:05 a.m. It remained within uninterruptedly until 9:00, or 115 minutes, then suddenly flew away. An hour and ten minutes passed without the arrival of either parent to resume incubation, and at 10:10 I left. The barbtails regularly entered their nest by flying swiftly toward it, low above the stream, to curve sharply upward until, ascending vertically, they shot into the downwardly directed doorway with no perceptible decrease in their speed. They left by darting out headfirst and flying rapidly away.

On 2 June I began to watch the nest at 5:15 a.m., while the light was still dim. At 5:27 a barbtail arrived, perched crosswise on a twig above the stream, called a few times with slight, high-pitched notes, then flew up into the nest. At 5:31 the other member of the pair flew into the nest and either replaced its mate or left immediately on finding the nest occupied. At 5:42 the barbtail who had remained within darted forth and flew upstream. At 5:46 one entered the nest. At 7:14 the other parent arrived and clung to a nearly upright slender branch about five yards from the nest. Although I heard no call, the incubating partner evidently sensed its presence, for it now left, after sitting for 88 minutes. The other promptly entered, but after incubating for only 28 minutes it flew off, at 7:42, without waiting for relief. Then I continued to watch for a long while without seeing either member of the pair return. Sessions of irregular length and long intervals of neglect are typical of incubating ovenbirds.

Two hours after the last barbtail went off leaving its eggs unattended,

I suddenly noticed a *zopilota,* a large, shiny, black snake, about seven or eight feet long, stretched out in a patch of sunshine on the log above the nest. It had slid down stealthily from the farther bank, which was higher, and molded its curves so well to the irregularities of the log that it was difficult to detect; it may have been lying there a good while before I noticed it. As I approached, the serpent crept back to the higher bank. This snake (*Clelia* sp.) is not known to plunder birds' nests but is reputed to prey largely on other snakes; in Brazil, where it is known as the *musarama,* it is, I am told, protected by law for this reason. Accordingly, I did not kill it to protect the barbtails' nest, but drove it away, although, being a bold snake that often stands its ground in the face of a man, it was reluctant to go. After its departure, I watched for another half-hour, or until 10:45; but still no barbtail arrived to warm the eggs, which had now been neglected for over three hours.

When I returned a few days later, the nest was empty but intact. I believe that only a slender snake could have removed the eggs without damaging the structure, but I have no assurance that the big *zopilota* which I had spared was not the culprit.

Spotted Barbtails roost singly in nests similar to those in which they breed. On my recent visit to Venezuela, Paul Schwartz showed me two mossy nests attached, not far apart, to the uneven rock-face of a vertical highway cutting, in the heavy, humid forest of the coastal range at Rancho Grande. These nests had the same form as the breeding nest of this species that I found in Costa Rica but were less massive, and the platform near the top seemed too slight to hold an egg. On a nocturnal visit, Schwartz had found a barbtail sleeping in one of these eggless nests, with tail downward and head inward. By means of banding, he proved that three barbtails successively occupied the same dormitory, always roosting singly. After being banded, the first barbtail moved to the neighboring nest, where at intervals over a period of a year Schwartz found it sleeping. On some nights it was inexplicably absent. No egg ever appeared in these nests. It was surprising to find these retiring forest birds roosting close beside a paved highway along which motorcars passed day and night.

Other species of ovenbirds, including the Red-faced Spinetail and the Plain Xenops likewise roost singly in dormitories similar to their breeding nests. Rufous-fronted Thornbirds are more sociable, and as many as six have been found sleeping in the same massive nest of sticks, hanging conspicuously from a tree.

Family PIPRIDAE

WHITE-RUFFED MANAKIN

Corapipo leucorrhoa

The White-ruffed Manakin is a stout, short-necked, short-tailed bird slightly over three and a half inches in length. The male is nearly everywhere glossy blue-black, with a gleaming white gorget that extends from his chin and throat around the sides of his neck and can be expanded as a ruff. The far plainer female is generally grayish olive-green, grayer on the throat, and becoming more yellowish on the abdomen and under tail coverts. In both sexes, the small bill is blackish; the eyes brown; the legs and toes dusky.

Young males in transitional plumage are far from rare, even in the nesting season. The first sign of the change to the adult colors is the whitening of the throat of birds in grayish olive plumage, like that of the females. According to Ridgway (1907:755), this whitening begins on the sides of the throat and advances toward the center. After the throat has become white—but still with a decidedly grayish cast—the sides of the head blacken, from the lores to behind the eyes. The next step is sometimes the blackening of the crown; but in other individuals black appears at the posterior margin of the white throat, and irregularly on the breast, while the crown is still largely olive. I saw a young male at this stage at the end of April; and in mid-June I found another male whose whole body was irregularly pied with black and olive, the black predominating. The males evidently do not acquire the full black-and-white adult plumage until they are over a year of age, if not older. The grayish white throat feathers of the male in transitional plumage can be erected as a modest ruff, and such males engage in courtship displays, as we shall see in due course. I have far more rarely seen males of the Blue-crowned and Yellow-thighed Manakins in transitional plumage, and scarcely ever those of the Orange-collared Manakin, even where this species is very abundant. In these three manakins, either the transition to the adult plumage is more quickly effected, or transitional males remain well hidden.

The White-ruffed Manakin ranges from Nicaragua to Colombia and Venezuela. On the Pacific slope of southern Costa Rica, it is found from sea level up to at least 5,000 feet, and it has been recorded at an equally high altitude in neighboring parts of Panama (Ridgway, 1907:755). Of the four small manakins in this region, the White-ruffed extends the highest. As one climbs upward from the Pacific lowlands of southern Costa Rica, the Orange-collared and the Yellow-thighed Manakins, abundant up to about 3,000 feet, drop out in the next 500 feet or so. I have traced the tiny Blue-crowned Manakin up to about

4,000 feet, where it becomes rare, just at the altitude where the White-ruffed Manakin is one of the most common birds of the forest. On the Pacific slope of the Cordillera de Talamanca, only the big, brown Thrush-like Manakin accompanies the White-ruff up to 5,000 feet. Where the four small manakins occur together, the males are distinguishable at a glance but the identification of the females is more difficult. The female Orange-collared Manakin is recognized by her orange or pinkish legs; the female Blue-crowned Manakin by her dull parrot-green plumage and her oft-repeated soft trill. To distinguish the female Yellow-thighed from the female White-ruffed is more difficult, as the two are olive-green birds of about the same size, but the latter is decidedly grayer.

Apparently the White-ruffed Manakins move seasonally up and down the mountains. On our farm at 2,500 feet in the valley of El General, they are by no means so constantly present as the other three small manakins. They are most in evidence from December to March. Although they sometimes remain into the nesting season in April, May, and June, I have seen no indication that they breed here; I have discovered none of the courtship gatherings that were not difficult to find in the forests near Cañas Gordas, 1,000 to 1,500 feet higher. White-ruffed Manakins were not rare in the coastal forests around the Golfo Dulce in the last quarter of the year, but I am not certain whether they stay to nest in this region. In the Sarapiquí lowlands on the other side of the country, (Slud, 1960:101) found White-ruffed Manakins from the end of November to March; during the rest of the year he recorded their presence only once in July and once in August.

White-ruffed Manakins live chiefly in the heavy forest, whence they sometimes venture into neighboring clearings with scattered trees and shrubs in search of food. When not engaged in courtship displays, they usually remain at moderate heights, from 15 to 30 or 40 feet above the ground. When they inhabit the same forests as the Blue-crowned, Yellow-thighed, and Thrush-like Manakins, vertical stratification of these species is evident, although of course the birds are not strictly confined to any level of the woodland. Lowest of all is the largest and dullest, the brown Thrush-like Manakin with large, dreamy eyes, who repeats his lovely tripartite whistle while clinging to an upright stem usually within two yards of the ground. Next come the Blue-crowned Manakins, which are mostly found from 6 to 15 feet up. Above these is the level where the White-ruffed Manakins are most often seen. Highest of all are the Yellow-thighed Manakins, which in El General display on such elevated perches that they are most difficult to observe; although in other regions, where fewer kinds of manakins are present, they are frequently found lower (Skutch, 1949).

White-ruffed Manakins wander through the forest in small parties which often consist of two or three adult males, several olive birds which may be either females or young males, and frequently also a male or two in transitional plumage. These little flocks draw attention to themselves by slight, thin, sharp notes which often sound like *see* and sometimes are almost little squeals. If one looks upward into the crowns of the smaller trees, he will see the black and olive birds flitting restlessly back and forth from twig to twig. The adult males often fluff out their white gorgets into a prominent ruff, and the immature males may do the same with their shorter, grayish white throat feathers. The males spread and flutter their wings, or rapidly open and close both wings simultaneously; but the watcher fails to hear the snapping sounds which familiarity with *Manacus* and *Pipra mentalis* leads him to associate with such gestures. Soon the restless band drifts away to a neighboring tree, whence the weak, sharp squeals continue to issue, although the dense foliage of the underwood makes the birds' movements difficult to follow with the eyes. Until I discovered the logs where White-ruffed Manakins display, I supposed that these drifting bands were courtship gatherings. The males in these groups do indeed give some of their courtship displays in a subdued form; but these animated parties of both sexes and varied ages are sometimes found in localities, such as on our farm in El General, where I have discovered no evidence of breeding.

The dietary habits of the White-ruffed Manakin seem to differ little from those of *Pipra* and *Manacus.* It eats many berries, especially those of shrubby and arborescent melastomes, which it frequently plucks on the wing. It also picks insects and spiders from the foliage in the same manner. These manakins are sometimes attracted by army ants and the attendant party of small birds of various kinds; but the White-ruffed Manakins that I have watched on the outskirts of the mixed flock of ant followers have seemed more interested in each other than in catching the insects which the ants stirred up.

Courtship

Although not so elaborate as the courtship performances of certain other manakins, those of the white-ruffed species are second to none in charm, and they have the advantage of being easy to watch, as these manakins seem indifferent to a human observer standing or sitting quietly a few yards away. The display centers about a smooth mossy log in the midst of the forest. The first display log that I found was slender, only five inches in diameter, and instead of lying flat it was slightly inclined, its upper end resting a few feet above the ground. The second display log was about 50 feet long and 15 inches in diameter, and it lay on a moderate slope, so that it also slanted. The third was a massive log about 18 inches in diameter, far advanced in decay and well covered with green moss, lying horizontally on the back of a forested ridge. The fourth display log was a large trunk that bridged a hollow in a forested slope, with its middle portion raised well above the ground. Three of these logs were in parts of the woodland with rather open undergrowth, but one was closely surrounded by low, dense vegetation that impeded the manakin's display.

Although each little circle of bare ground is the display court of a single male *Manacus* and each horizontal display branch belongs to a single Yellow-thighed Manakin, the mossy display log is used by several White-ruffed Manakins simultaneously. The three logs that I watched most carefully belonged to three, three, and four adult males, and two of these logs were also the scenes of displays by one or more young males in transitional plumage.

While walking along a forest path soon after sunrise on a morning in early April, I noticed several adult male White-ruffed Manakins ahead of me. While one chased another, a third gave a beautiful display, such as I had never before seen. Flying slowly and bouncing up and down, with the snow-white feathers of his ruff widely spread and contrasting prettily with his blue-black plumage, he crossed the path obliquely, to alight on the slender mossy log already mentioned. Standing on this log, facing up its slope, with his throat feathers still spread horizontally to their fullest extent, he lowered his foreparts and bent down his head, as though attentively examining some minute object on the mossy surface. After a few moments in this posture, he flew up into the surrounding bushes.

I stood enchanted while this charming display was repeated several times, by the same individual or others. But it required repeated observations during the following days to become familiar with all the details of this unique performance. The bouncing flight, in which the manakin traces rounded peaks and valleys in the air, is typically begun at a point amid the undergrowth 15 or 20 feet from the display log and on any side of it—there is no fixed starting point. The strongly undulatory course inclines gently downward to the log. With his wings beating slowly through wide arcs, the bird advances slowly, sometimes seeming just to avoid the stalling point. His little body is held nearly upright; his tail is raised until it is almost parallel with his nearly vertical back; his plumage is puffed out making him appear quite round: he resembles a tiny black balloon with a gleaming white patch on its forward side, just below its top, as he bounces through the air toward his mossy landing platform. One manakin may perform this flight while the others are out of sight; or the whole group may be present, one or two resting on the log while another undulates toward it and one or two more fly around or perch nearby. Each, however, performs independently, without trying to coordinate his movements with those of his companions. At times, however, as many as three male manakins fly around their log simultaneously in slow, up-and-down flight, without directing their course to it. These slow, bouncing flights were always performed in silence.

In addition to the slow, undulatory, silent approach to the display log, there is also a rapid, direct, noisy one. This obliquely downward dart was often so sudden and swift that I first became aware of it when the manakin made a loud noise, evidently with his wings, just as he neared the log. The sound is hardly a firecracker-like *snap,* such as is made by *Manacus* and *Pipra mentalis,* but rather a duller *flap,* which may be roughly imitated by suddenly jerking taut a piece of stout cloth allowed to hang loosely between one's hands. The *flap* is often followed immediately by sharp, harsh, little notes, *cheee waaa,* which reminded me of the shrill vocal sound made by a male Blue-crowned Manakin as he alights on his nuptial perch in the undergrowth of the forest. Thus the full sequence of sounds, as the swiftly approaching manakin is about to alight on the log, is *flap cheee waaa;* but the vocal notes are sometimes omitted. Although usually this swift, noisy approach to the log begins at a point hidden from view by the foliage, at times a manakin flying toward it with slow undulations suddenly accelerates his progress and, with a swift jerky movement, makes the *flap,* followed by the *cheee waaa,* as he alights. Often a manakin chooses this boisterous mode of approach to come down beside another male resting on the log; and this startling arrival may, not surprisingly, send the other away. Or the displaying manakin may fly close above another on the log, making the *flap* as he passes overhead, and continuing onward to alight in the bushes beyond.

After alighting on the log, the displaying manakin usually, as I saw with the first one I watched, rests with his ruff widely spread, his foreparts depressed, and sometimes with his head turned sideways, as though to examine the log with one eye. While standing so, he often twitches his folded wings. He may maintain this posture for a few seconds. Sometimes he hops along the log, or flits from one part to another. If several males are on the log at the same time, they do not interfere with each other—the display log is long enough to accommodate, without crowding, the four or five little birds who use it and a good many more. After delaying a short while on the log, the manakin often flies away with the same slow, spectacular, bouncing flight that took him to it. Then, after flitting through the undergrowth, he may repeat the whole performance.

When not actively displaying, the male manakins spend much time, especially in the forenoon, perching in trees close by their mossy log. Here they settle on slender branches, usually from 15 to 30 feet above the ground, where, with their feathers all puffed out, they rest motionless a few feet or yards from each other, remaining so for many minutes with only an occasional shift of position. Adult and transitional males repose together in this fashion, seeming to enjoy each other's company. Sometimes they preen. From time to time one of the perching adults spreads and beats his wings to make a resonant *flap*, such as one hears as the manakin flies swiftly to the log. Although a single sharp stroke of both wings together produces this noise, I have watched perching manakins beat their wings almost too rapidly to be seen, without making a sound audible to me. After resting for a good while, the birds either renew their display or drift off through the forest.

Although the several males who display at the same log usually seem to preserve the friendliest relations, occasionally one chases another. Around and around they go, weaving through the undergrowth in the vicinity of the log, often continuing the rapid pursuit for a surprisingly long while. I never saw the pursuer overtake the fugitive; indeed, I have never known manakins of any kind to clutch together in a fight. These chases are accompanied by the slight, shrill notes that are so frequently heard from the mixed parties that flit through the trees as already described, but are never, in my experience, uttered by males actually displaying at their logs.

The only display log at which I found four adult males was the longest of all, about 50 feet in length. This prostrate trunk was so closely surrounded by undergrowth that the manakins could not approach it by means of the usual gently descending undulatory flight, but perforce dropped almost straight down to it, whether they arrived silently or with the loud *flap*. The small amount of bouncing flight that I saw here was chiefly along the length of this long log, in the clear space above it. At one time three adult manakins stood close together with lowered foreparts on this log, while a fourth flew nearby. With the three adults on the log was a manakin who resembled a female; but since the adult males ignored it, I inferred that it was a young male.

The young males that I saw displaying at other logs differed from females chiefly in their whitish gorgets. They might or might not show a little black on their plumage, chiefly on the cheeks and sometimes also at the posterior margin of the light throat patch. They displayed much as the adult males did, approaching the log in the butterfly-like bouncing flight and sometimes also in the swift flight accompanied by the *flap cheee waaa*. While standing on the log, they spread their half-developed ruffs much as the adults spread their full ones. These young males displayed either with the adults or while the latter were absent. The adults seemed indifferent to the presence of the young males. I did not see adult males pursue immature males, as they chased each other.

In addition to these displays at the logs, in early April I watched a White-ruffed Manakin with no sign of maleness in his plumage, but probably of this sex, performing all alone in the midst of the forest. For many minutes, he flew back and forth from twig to twig, in an erratic course which covered an area perhaps 50 feet in extent.

He remained low, from about 5 to 20 feet above the ground, always silent and alone. This behavior reminded me strongly of one of the displays of the male Blue-crowned Manakin but seemed unrelated to the other displays of the White-ruffed Manakin that I witnessed.

The activity at the display logs was at its height in April and continued through May, but after this it rapidly waned. In April the time of most active display was from around sunrise to about the middle of the forenoon, with sporadic bursts of displaying later in the day. Despite many hours of watching, I never saw an undoubted female visit a display log. The few times when a manakin in female plumage was present, the indifference of the males suggested that this was a young bird of their own sex. From analogy with other species, I have no doubt that the swift approach with the *flap cheee waaa* is used by the male as he comes to mount a receptive female.

In sharing the display stage rather than performing on one that is strictly individual, the White-ruffed Manakin resembles *Chiroxiphia* rather than *Manacus* and the Central American species of *Pipra*. But whereas two or more individuals of *Chiroxiphia* join in beautifully coordinated displays that have been described by various authors (Gilliard, 1959; Sick, 1959; Slud, 1957; Snow, 1956, 1963); the White-ruffed Manakins make no attempt to coordinate their activities at the display log which they share. Slud (1964:238) stated that the arrival of a second male on the log is the signal for the departure of the individual already there, so that never more than one is present on the display log at a time. But the manakins I watched did not show even this degree of attention to each other's movements, and two or three not infrequently stood on the log together. Possibly Slud watched the Caribbean race *altera*, whereas my study was confined to the Pacific race *heteroleuca*; this might account for our divergent observations.

A brief note by Davis (1949) on the display of the White-throated Manakin in British Guiana reveals some interesting similarities as well as differences between this species and the White-ruffed Manakin that it rather closely resembles in plumage. Davis watched a small mixed party of both sexes perching and chasing each other amid the branches from about 10 to 50 feet above the ground in heavy primary forest. Presently a female flew down to an old mossy log, where she was joined by a male, and coition occurred without any preliminary display. Later a male came to the log and crouched with his bill pointing upward, in which attitude both his white throat and a longitudinal bar of glossy steel blue on his wings were conspicuous. After this bird had left, a female came to the log, where she was joined by a male who, crouching and spreading his wings horizontally to their fullest extent to reveal white wing bars at the bases of the primaries, approached her with a slow and labored undulating crawl. Another male drove him off before he reached the female, who also flew away. In displaying on a mossy log, the White-throated Manakin resembles the White-ruffed Manakin; but the latter was never seen to crouch with its bill pointing straight upward—it displayed its white gorget in another manner—nor to crawl over the log.

Nesting

On 29 March 1964, I found a female building a nest in a deep forested valley near Cañas Gordas, at an altitude of about 3,500 feet. Her site was the horizontal fork of a slender twig of a small tree, among its outer foliage, 22 feet above the ground. Here she had already accumulated a thin, loose weft of fine, light-colored fibers, suspended hammock-like between the arms of the fork. My attention was drawn to this nest by seeing the manakin approach it with fibers in her bill, at 11:15 in the morning. In the next half-hour, she brought material five times. On the following morning, she continued to build slowly, visiting her nest six times between 6:05 and 7:00 a. m., five times in the following hour, and only three times from 8:00 to 9:00. On some of these visits, she merely deposited her material while standing beside the nest and promptly left; on others, she sat in the structure, shaping it, for periods up to 4¼ minutes. She turned from side to side, using her bill, and apparently also her feet, to arrange the materials. Her movements were slow and deliberate; she was invariably silent and alone.

When finished, this nest was a very shallow hammock, composed chiefly of long, round, smooth, rather stiff brown filaments which appeared to be fungal rhizomorphs. Mixed with these were some darker and some lighter strands of the same character. The bottom of the nest was partly covered with lacy fragments of leaves reduced to skeletons by decay and also a few long skeins of light-colored bast fibers. A little cobweb had been applied to the supporting arms, but not enough to attach the nest securely. Although the nest was fairly thick, much light passed through the bottom. The over-all measurements were about 2½ inches in diameter by 1¼ inches in height. The concavity that held the eggs was barely ½ inch deep.

By 7 April the nest contained two eggs, which I could see, without much disturbance of the surroundings, by raising a small mirror on the end of a very long, light pole and examining the images through my binoculars. Thus viewed, the eggs appeared dull white, heavily marked with brown, one over its whole surface, the other chiefly on the thicker end. The female seemed to take a long while to settle down to regular incubation; and before I could study this activity, her eggs vanished. Much searching through the neighboring forests failed to yield another nest. Indeed, the probability of discovering such a small nest well above one's head amid clustering foliage is remote, unless one happens to notice a bird carrying material to it, or bringing food for nestlings.

Family COTINGIDAE

BARRED BECARD

Pachyramphus versicolor

Only four and three-quarters inches in length, the Barred Becard is a stout little bird with the large head typical of its family. In the male, the crown, hindneck, back, and scapulars are glossy black. The rump and tail are gray, with narrow white tips on all but the central rectrices. The wings are black, conspicuously marked with white. The sides of the head and neck, the chin and throat are olive-yellow. The remaining under parts are dull white, more or less tinged with yellow on the breast, and finely barred with dusky except on the under tail coverts. The female has the pileum and hindneck slaty gray. Her back, rump, and upper tail coverts are plain olive-green and her tail is grayish brown. Her wings are dusky with conspicuous areas of cinnamon-rufous on the coverts and broad light edgings on the remiges. The sides of her head and neck are yellowish olive with pale yellowish eye rings. Her ventral plumage is barred much as in the male but more strongly tinged with yellow. In both sexes, the short and rather thick bill is blackish, as are the legs and toes.

The Barred Becard ranges through the highlands from Costa Rica to Bolivia and western Venezuela. In Costa Rica I have found it only between 5,000 and 7,500 feet above sea level, in the vicinity of Los Cartagos and Vara Blanca, where it was rare, and farther east on the same volcanic range, on the northern slope of Volcán Irazú, where it seemed somewhat more abundant. Doubtless the bird lives both below and above these limits, but published statements of its altitudinal range are vague. According to Phelps and Phelps, Jr. (1963: 144), in Venezuela the species is found from about 6,500 to 9,500 feet above sea level.

The Barred Becard roams through the cool, mossy mountain forests, often in mixed flocks that include ovenbirds, woodcreepers, flycatchers, wood warblers, vireos, and other small birds. Usually it stays high in the great, epiphyte-covered trees, but occasionally it descends into the underwood, and Slud (1964:229) has seen it venture forth into low shrubbery beyond the woodland. I have seen too little of this becard to determine whether it remains paired after the close of the breeding season. It subsists largely on insects which it catches as it darts against the foliage. Its restless activity and the briskness of its movements contrast with the more deliberate actions of its larger relatives.

A pair of Barred Becards that I watched while they built their nest often repeated a rather high-pitched, soft, endearing *weet weet weet weet weet* that was sometimes long-continued. The cozy notes reminded me of the Tufted Flycatcher's voice. These notes differed in pitch, and I watched the male deliver a series of *weet*'s deeper than most that I heard. Occasionally the sequence rose in pitch and became faster toward the end. Another male uttered a high, weak trill, scarcely audible above the incessant roar of the mountain torrent that rushed through the deep gorge where he and his mate were attending nestlings.

Nesting

On 22 June 1963, I found a pair of Barred Becards building in heavy forest of oaks and other broadleafed trees with an undergrowth of tall, cane-like bamboo, at an altitude of 7,500 feet. The site they had chosen for their nest was about 75 feet above the ground, on a thin, leafy branch at the top of a tall, slender tree standing in a small opening in the forest, with its narrow crown isolated from other trees. Their nest, already far advanced, was a great globular structure which seemed to be about one foot in diameter. It was composed, at least externally, largely of green moss, strands of which hung beneath it and swayed gracefully in the breeze. When I first noticed this nest, both sexes were carrying material to it; but they soon left, and I waited long without seeing them again.

Ten days later, I again found the pair of becards working at their nest, and for three-quarters of an hour they continued to build actively. They brought tufts of green moss for the top and dry bamboo leaves to take inside. Both sexes brought both kinds of materials, the male sometimes flying up with a tuft of moss longer than himself. Apparently, in contrast to a pair of Rose-throated Becards that I once watched while they built, the male as well as the female entered this nest; but since the doorway faced down the steep slope, away from me, I could not make sure of this. As they laboriously gathered material, these becards repeated the soft, appealing calls that I have already described.

I do not know whether eggs were ever laid in this high nest. The lateness of the season, and the long time that the becards took to complete their structure, made me doubt that they would be. Possibly these birds were engaged in post-breeding building activity, such as one sometimes finds in castlebuilders (*Synallaxis*). Possibly, too, they were building a dormitory nest, but the site was so far from my dwelling that I could not investigate this point.

Many years earlier I had found a Barred Becard's nest with young. It was situated about 50 feet up, in the very top of a slender tree that had a clean trunk and stood apart from its neighbors in the profound gorge of the Río Sarapiquí, at an altitude of about 5,100 feet. This was also a bulky, roughly globular structure that seemed to be about a foot in diameter and, as far as I could see, was composed chiefly of green moss and slender lengths of dead herbaceous vines. It rested between four thin, ascending branchlets which diverged from the end of a slender, upright branch. The four points of attachment broke the generally globular form of the nest, which projected a short distance up each supporting branch. By its form and covering of green moss, this nest, like that which I afterward watched the becards build, might have been taken for the work of the Red-faced Spinetail, whose altitudinal range overlaps that of the Barred Becard. But the spinetail's nest usually swings at the end of a dangling vine or a thin drooping branch of a tree, whereas the becards' nests were upheld from below. It is noteworthy that both were in trees with clean, slender trunks and boughs which did not interlock with those of neighboring trees, thus making these conspicuous structures more difficult for climbing animals to reach.

When I found the nest in the gorge of the Río Sarapiquí on 4 May 1938, both parents were carrying food to the nestlings. Their movements showed me the position of the doorway, which without this help I should never have distinguished from the ground. It was in the bottom of the nest, well concealed among the dangling tufts of moss. Now, in the middle of the morning, the nestlings did not require brooding, and a moment after a parent darted into the nest with a billful of food it shot out again. If the young in the nest were as loquacious as nestling White-winged Becards, their infantine voices were overwhelmed by the torrent's clamor, which filled all the wild gorge where they were hatched.

Family TYRANNIDAE

DARK PEWEE

Contopus lugubris

This pewee resident in tropical highlands is distinctly larger and darker than the migratory wood pewees familiar in the north, and it has a much more pronounced crest. About six and a quarter inches long, it is nearly everywhere dark sooty gray, with paler gray on the throat and yellowish white on the center of the abdomen and under tail coverts. The dusky plumage is also relieved by the narrow white edges of the secondaries. The upper mandible is blackish and the lower yellowish; the legs and toes blackish. The sexes are alike.

Confined to the highlands of Costa Rica and adjacent parts of Panama, the Dark Pewee lives chiefly between 4,000 and 7,000 feet above sea level. The only places where I found it common were on the northern slope of the Cordillera Central in the vicinity of Vara Blanca, between 5,000 and 6,000 feet, and farther east on the same volcanic range, on the northern face of Irazú at corresponding elevations. Although closely associated with the tall, epiphyte-laden forests of these rain-beaten mountain slopes, the Dark Pewee prefers their edges and openings in their midst to their unbroken interior, and it is often found in adjoining clearings with scattered trees. Like the related (perhaps conspecific) Greater Pewee in Guatemala, the Dark Pewee separates from its mate at the close of the breeding season and through much of the year avoids others of its kind. But the Greater Pewee generally joins a mixed flock of warblers and other small birds which roam the woods in the winter months, one individual to a flock; whereas the Dark Pewee prefers to lead a wholly solitary life during the latter part of the year. Not being attached to a wandering flock, it is also more stationary, as befits its manner of foraging.

Each Dark Pewee has a favorite resort, where it is to be found alone day after day. Perhaps most often the flycatcher establishes its headquarters at the edge of the forest, where it perches on a high exposed twig, preferably of a dead branch, and not infrequently at the very top of the tree. From this lofty watchtower it sallies forth to snatch up passing insects, returning after each long foray to its favorite perch. Sometimes it catches butterflies of fair size, which it gulps down wings and all. In its large size, dark coloration, and preference for high perches for flycatching, it resembles an Olive-sided Flycatcher more than one of the smaller wood pewees.

Voice

While perching upright, keeping a sharp lookout for passing insects, the Dark Pewee advertises its presence by a rather loud, staccato *wic,* which is usually repeated several times in succession. This characteristic utterance is heard with frequency at all hours of the day. It is quite similar to the call note of the Dark Pewee's northern relative, the Greater Pewee. Far more rarely than the sharp *wic,* the Dark Pewee gives voice to a short, low-pitched, fairly loud whistle. In April and early May, this flycatcher mounts at daybreak to a perch on a high treetop to deliver a distinctive dawn song. *Fred'rick fear,* he sings, *Fred'rick fear fear,* over and over for many minutes together. The name *Fred'rick* is pronounced quite clearly, with the last syllable ascending in pitch, and in a higher key than the *fear* which follows it. This *fear* is the above-mentioned fairly loud whistle that is not restricted to the dawn song but is heard even outside the nesting season, although far more rarely than the *wic.*

The full *Fred'rick fear* is seldom given except at daybreak in the breeding season but, as is true of the dawn songs of many kinds of birds, it may be delivered at any time of day by one who is highly excited. Indeed, I first heard this song late on the sunny morning of 19 March, when a pewee in a treetop repeated it over and over, interspersed with many staccato *wic*'s. The stimulus for this untimely outburst of dawn singing was evidently another pewee who perched in a neighboring tree, calling *wic wic wic* but never uttering *Fred'rick fear,* whence I inferred that it was a female. At this season the Dark Pewees, who had been solitary earlier in the year, seemed to be forming pairs.

The discovery of the Dark Pewee's dawn song solved a mystery of long standing. Five years earlier, among the pine-oak woods around 8,500 feet in the Guatemalan highlands, I had heard a very similar song on a number of mornings in the breeding season, but always in the dim light of dawn, by a bird who persistently stayed so high in the treetops that I never glimpsed him. I called this elusive songster, which I felt certain was some kind of flycatcher, the "Frederick Fear Bird," and later I heard its song in the pine woods on the Pacific slope of Guatemala, at an altitude of about 3,000 feet. After identifying the dawn song of the Dark Pewee, I had little doubt that my mysterious "Frederick Fear Bird" was the closely related Greater Pewee, which is widespread in the Guatemalan highlands. The resemblance of the most common call notes of these two birds has already been mentioned.

Nesting

At Vara Blanca in 1938, I found three nests of the Dark Pewee, all around 5,500 feet above sea level. The first was discovered on 28 April, when it already held one egg. This nest was attached to the upper side of a slightly descending branch of an *Erythrina* tree growing in a pasture. It was near the end of the bough, between three ascending branchlets, at a height of 16 feet above the ground. The second nest, found on 7 May, was near the upturned end of a descending limb of a small, slender tree growing just within the edge of the forest that filled the great gorge of the Río Sarapiquí. This nest was saddled over the supporting branch, among the terminal branchlets, by the foliage of which it was shaded. The tree stood on a slope so steep that to give the nest's height above the ground would be meaningless. I could not reach this nest, but evidently it contained eggs which a pewee was incubating. The third nest was about 40 feet up in the top of a small tree growing in a pasture, in front of a Quetzal's nest that I was studying. This pewees' nest contained nestlings, which both parents were feeding, on 28 June.

The first nest, found with one egg on 28 April, never received another, as far as I could tell. After it had remained unattended for a number of days, I took the nest down for closer scrutiny. It was a broad, shallow cup that fitted like a saddle over the moderately thick branch and did not depend for support on the ascending branchlets amid which it was placed. The massive outer wall was composed of green mosses, liverworts, and lichens, including the foliaceous *Cora,* all bound together with cobweb. The cup was thickly lined with fibrous rootlets and coarse vegetable fibers. The cavity of the nest measured $2\frac{1}{2}$ by $2\frac{1}{4}$ inches in diameter by 1 3/8 inches deep. A beautiful structure!

To judge by the behavior of the Dark Pewees themselves, their chief enemy in the nesting season is the Blue-throated Toucanet, abundant at the altitudes where these flycatchers live. When one of the great-billed green birds alights anywhere near a pewees' nest, the two parents become intensely excited, calling loudly as they dart back and forth close above the intruder, but never daring to come within reach of the highly colored, menacing bill of the far larger bird. Only when the toucanet, at its own good time, starts to fly away, do the pewees dart close enough to pull its feathers. The Dark Pewee's antipathy to the toucanets is as strong as that of the bigger Boat-billed Flycatcher to the far larger Chestnut-mandibled Toucans of the lowlands, or the Tropical Kingbird's enmity to the Swallow-tailed Kite.

Since some systematists regard the Smoke-colored Pewee of the South American mountains as conspecific with the Dark Pewee and the Greater Pewee, my few observations on this form may not be out of place here. In eastern Ecuador in 1939, I found Smoke-colored Pewees from about 4,000 feet in the Pastaza Valley up to at least 7,500 feet on Volcán Tungurahua. Their dark gray plumage resembled that of their Costa Rican relatives, but their two obscure light wing bars set them off as somewhat different. Their crests were less pronounced than in the northern birds. Their call, *wic wic wic,* was very much like that of the Dark Pewee and likewise that of the Greater Pewee in northern Central America. At an altitude of 7,400 feet in the great barranco at the foot of Tungurahua, I found a nest on 12 October. It was a shallow mossy cup, built in the fork of a mossy branch, about 50 feet up in the top of a small tree that stood at the edge of a clearing. On this date there were two nearly feathered nestlings, which were fed by both parents and still brooded for very brief periods. Nine days later the nestlings, now fully feathered, rested on the rim of their nest, which had become too small for them.

YELLOWISH FLYCATCHER

Empidonax flavescens

The Yellowish Flycatcher is an easily recognized member of a difficult genus. It is nearly five inches in length, deep yellowish olive-green on the upper parts, with grayish brown wings and tail. There are two conspicuous bars of cinnamon-buff or greenish buff on the wings, the larger feathers of which have lighter margins, as do the tail feathers. Behind each dark eye is a conspicuous pale yellowish or whitish crescent, and before the eye is a narrower crescent of the same color. The under parts of the body are yellow, more or less tinged with olive, especially on the chest and sides. The upper mandible is blackish, the lower yellowish, and the legs and toes are dark.

This little flycatcher is resident through the mountains from southern Mexico to western Panama. In the former country it has been recorded at altitudes ranging from about 3,000 to 10,000 feet, and in Guatemala I found it up to nearly 10,000 feet. Strangely enough, at the southern end of its range it remains lower and has not been recorded much above 7,000 feet, at which altitude the congeneric but very distinct Black-capped Flycatcher of the high mountains makes its appearance. On the Pacific slopes of the Cordillera de Talamanca, the Yellowish Flycatcher hardly descends as low as 4,000 feet; but in cool pockets on the Caribbean slope of Costa Rica it is occasionally found at, or even below, 2,000 feet (Carriker, 1910:698). Its zone of greatest abundance in Costa Rica is between 5,000 and 6,500 feet.

The Yellowish Flycatcher is found in the cool, damp mountain forests, especially in openings and around their edges, but it also inhabits open woodland, isolated patches of second-growth, and even highland pastures with scattered shade trees. Except while singing at dawn, it seldom ascends high above the ground. A solitary bird, it rarely joins in mixed flocks, and it even separates from its mate after the breeding season; at Vara Blanca, where it was common, I nearly always found it alone during the second half of the year. In its unsociable temper it resembles its northern congeners which winter in Costa Rica, especially the Yellow-bellied Flycatcher. But it is more confiding in the presence of man, often appearing quite indifferent to his close approach, and attending its eggs or young while he watches nearby, unconcealed. It catches flying insects on short sallies into the air, and often it plucks an insect or a spider from the trunk of a tree, beside which it hovers. Sometimes it drops momentarily to the bare ground beneath trees, or to the grass in pastures, to pick up food.

Voice

Early in April of 1963, I found a Yellowish Flycatcher singing at daybreak in a large, spreading alder tree growing in a mountain pasture at an altitude of nearly 7,000 feet on the Barba massif in Costa Rica. He performed well up in the tree, 30 or 40 feet above the ground. Beginning while the dawn light was still too dim to reveal his colors, he repeated interminably a phrase that sounded like *seee seee chít*, the first two notes weak and sibilant, the third sharper. This verse was uttered about 21 times per minute, with scarcely a pause between one repetition and the next. The performance was continued almost without interruption for a quarter of an hour or more, until the light grew stronger. Before sunrise, the flycatcher delivered his last sharp note and flew across the pasture to a row of cypress trees.

Throughout April and well into May, the Yellowish Flycatcher continued to perform at daybreak in the same alder tree. One morning toward the end of April, I found that he had slightly varied his tune, now singing *seee chít seee*. But a few days later he had returned to his old phrase *seee seee chít*. While this flycatcher proclaimed his presence in his usual place on 10 May, another was singing a similar refrain in a neighboring tree. One may question the propriety of designating as song an utterance so simple as that which the Yellowish Flycatcher monotonously reiterates at dawn. However, it seems to merit this designation if one accepts the criterion of Armstrong (1963:41) that "song may be considered to constitute the major item (or items) in a bird's vocal and/or instrumental communicatory repertoire, and normally the most informative, complex and sustained." Intensive dawn singing, comparable to that of the Yellowish Flycatcher, has been recorded for a number of northern species of *Empidonax*, including the Least Flycatcher (MacQueen, 1950:200-202), Western Flycatcher (Davis *et al.*, 1963:346), Gray Flycatcher (Johnson, 1963:157-159), Acadian Flycatcher (Mumford, 1964:14-15), and others. In the prolonged twilight at midsummer at high latitudes, some of these little flycatchers continue their dawn singing far longer, as well as at a more rapid pace, than does the Yellowish Flycatcher.

The common call note of the male Yellowish Flycatcher is a *seee*, similar to one of the notes of his dawn song. The female utters a weaker *seee*.

Nesting

In the Costa Rican highlands at 6,500-7,000 feet in 1963, nesting began not later than March, for on 23 April I found young already fledged. Two nests held eggs in May, and in the latest of the three that I discovered, laying began on 11 June. In a narrow ravine on the Sierra de Tecpam in west-central Guatemala, at about 8,500 feet, I found a nest with young on 11 July 1933. This mossy open cup was difficult to distinguish from the moss-covered cliff on which it was built, in a cranny close by a little waterfall, and it was further concealed by the large fronds of a chain-fern that draped in front of the niche. Not only were these flycatchers breeding in the middle of the rainy season of these cool heights, they had chosen a very wet situation for their nest. Even on a sunny day water seeped down the rock against which the back of the nest rested. Spray from the waterfall fell just short of it and probably reached it when the current was swollen by rain.

Of the three Costa Rican nests that I have seen, one was 6 feet above the ground in a deep fissure in the side of a living cypress tree, in an open grove of these trees amid pastures. The second nest was 15 feet up in an open niche in the side of a large, spreading, epiphyte-laden tree standing alone in a pasture, not far from a small patch of woods. Both of these nests were built with their outermost portion even with the surface of the supporting trunk, so that part of the nest, and the head of the incubating flycatcher, were visible from the ground; the nests were not in dark crannies but in fairly well illuminated situations. The third Costa Rican nest was 4 feet above the base of a vertical cut bank 10 feet high, beside a little-used roadway between pastures. The niche in which this nest was set was so shallow that much of the structure projected beyond the surface of the rather bare earthen bank, but overhanging grasses kept it dry even during the hard rains of this season.

All of these nests, in Guatemala and Costa Rica, were bulky masses of green moss, or of mosses and liverworts, interlaced with a few fibrous rootlets, horsehairs, or similar binding materials. In the Guatemalan nest, a few dead pine needles lined the open cup. The Costa Rican nests were deep cups lined with more varied materials, including rootlets, vegetable fibers, black or white horsehairs, a few fragments of papery bark and grass blades, and (in two nests) capsule stalks of the moss *Funaria* with the spore cases still attached.

When found on 11 July, the Guatemalan nest held two nestlings with expanding plumage and one unhatched egg. The earliest of the Costa Rican nests contained three eggs when I first saw it on 26 April; the second, an undetermined number of nestlings on 7 May. The third nest had one egg on 11 June and two on the following day. When I next visited this nest it had been destroyed, and I could not learn whether a third egg was laid. In the set of three, the short-ovate eggs were dull white, speckled and blotched with pale brownish cinnamon, thickly on the broader end, more sparingly elsewhere. These eggs measured 17.5 by 14.2, 17.6 by 14.0, and 17.1 by 14.0 millimeters. The unhatched Guatemalan egg, which measured 17.5 by 13.5 mm., was marked with rusty brown in a similar pattern. A nest found by Carriker (1910:698) at Juan Viñas, Costa Rica, on 8 May 1907, "was placed on the broken and jagged end of a fallen tree, about ten feet from the ground, in the midst of the forest." In construction this nest resembled those already described, and it contained three eggs, each of which measured 19 by 14 mm.

I watched the nest in the fissure in the trunk of a cypress tree throughout the forenoon of 29 April, a damp, chilly morning, with mist and drizzle driven through the trees by a moderate north wind and, after 10:30, light intermittent showers. Only the female incubated, taking 12 sessions which ranged from 9 to 38 minutes in length and averaged 20 minutes. Her 12 recesses ranged from 5 to 15 minutes and averaged 9 minutes. She covered her eggs for 68 per cent of the 6 hours that I watched. Usually she sat with her tail against the rear of the fissure and her head outward, just visible to me above her nest's mossy rim; but occasionally she sat with her tail outward, and still more rarely she rested sideways in her nest, with her head turned to look out.

The male flycatcher neither shared incubation nor fed his incubating mate, but he spent much time in view of the nest, perching on low branches of the nest tree or on the fence wires that were fastened to its trunk. He foraged much around the nest, snatching insects from the air, the bark of trees, or the ground. He chased a Spotted-crowned Woodcreeper from the trunk that held the nest and from a neighboring tree. Once his mate drove him from near the nest, then entered to resume incubation. Later, in her absence, he went six times in succession to look into the nest, sometimes while clinging to the side of the fissure just above it and sometimes while hovering in front. While inspecting the eggs, he uttered very low, soft notes.

The nestlings are fed by both parents. The two that hatched in the cypress tree stayed in their nest for 17 days. On the morning of 23 April 1963, I found two stubby-tailed fledglings amid dense shrubbery at the forest's edge, at La Giralda. They rested on a slender twig slightly above my head, in contact with each other and facing the same way, preening much and from time to time stretching a wing. The two parents, busily bringing them insects, were fearless of me, and one of them, on its way to the nestlings, rested hardly more than arm's length from me. For more than a quarter of an hour, the young stayed in the same position, always pressed close together, as one often finds juvenile flycatchers of various kinds.

TUFTED FLYCATCHER

Mitrephanes phaeocercus

This engaging flycatcher is about four and a half inches long, and the sexes are alike. In contrast to many other members of its family, it is easily recognized—by its high, upstanding crest, resembling that of the Tufted Titmouse, and the tawny-cinnamon or dull orange that forms a half-collar on the throat and sides of the neck and extends over the breast. The upper plumage is bright greenish olive, becoming grayish brown on the upper tail coverts and tail. The dusky wings have two olive or buffy bars on the coverts. The under plumage, posterior to the breast, is pale buffy yellow. This description is of the race *M. phaeocercus aurantiiventris* of Costa Rica and western Panama, with which the present account deals.

The species ranges from northwestern Mexico to Peru and Bolivia. In Mexico it nests chiefly between 6,000 and 10,000 feet, and in the northern part of the country it descends to the tropical lowlands in winter (Miller *et al.*, 1957:95-96). In Guatemala, where it lives between 5,000 and 9,500 feet above sea level, it is, at least on some of the higher ranges, the most abundant and conspicuous member of the flycatcher family. In Costa Rica, the Tufted Flycatcher remains lower and is rarely found above 7,000 feet. On the Caribbean slope, it extends downward locally to 2,000 or even 1,500 feet but it is not abundant below 3,000 feet. On the drier, more sheltered Pacific slope of southern Costa Rica, it hardly descends below 3,500 feet. Since I earlier (1960) published observations on the Tufted Flycatcher in Guatemala and northern Costa Rica, I shall limit this account to a recent study made in the Cañas Gordas region near the Panamanian border, where the species was moderately abundant between 3,500 and 4,000 feet.

In this region, the Tufted Flycatchers dwelt in the heavy forest, usually at points where the fall of a great tree had left a gap in the high canopy through which more light penetrated to the lower levels. Here they were present in the same small area day after day, nearly always in pairs, as in northern Costa Rica and Guatemala where I had found these flycatchers throughout the year. Perching well above a man's head but far below the crests of the tallest trees, usually between 12 and 40 feet above the ground, the sprightly little birds made frequent sallies to capture tiny flying insects, returning again and again to the same exposed lookout. At intervals they repeated their high-pitched, endearing *chee chee chee chee*—the series sometimes consisting of as many as seven similar notes, becoming more widely spaced toward the end. I failed to hear the indescribably quaint dawn song which I had heard at Montaña Azul many years earlier. *Bip-bip-bip-dididiidup-bip-bip-bibibiseer* the flycatcher seemed to sing in a thin, high voice, pouring out the syllables faster than human lips can form them.

Nest Building

On 9 April 1964, I found a nest under construction 50 feet up on a liana hanging above a forest path. The nest rested on a horizontal length of a dangling loop of the woody vine, about half as thick as the structure was wide, beside a long, hanging dead liana that crossed the loop. The supporting vine was overgrown with moss, and the exterior of the nest was made of green moss which the builder gathered from neighboring boughs.

On the following day, I started to watch this nest at 6:15 a.m., but the pair of flycatchers did not arrive until 7:40. In the next 20 minutes, the builder carried material to the nest 20 times. Then the pair went beyond sight and were absent until 8:50. At this hour one member, evidently the male, arrived and for a few seconds sat in the nest, twittering softly. After he flew to a neighboring perch, his mate resumed work. From 8:50 to 9:17 she made 24 trips to the nest. From 9:17 to 9:26 she did not work. From 9:26 to 10:00 she came to the nest, usually if not always with material in her bill, 32 times.

As far as I could tell, the male never helped to build. Much of the time he perched on the vine near the nest, whence he made frequent sallies to catch small flying insects. Often he repeated his *chee chee chee chee chee* or *chee chee chee chee chee chee* (the second trio of notes more widely spaced). From time to time he uttered a thin, high-pitched *seer,* like the final note of the dawn song. At a nest built, in May of 1938, 80 or 90 feet up in the top of a tall tree at Vara Blanca, the pair behaved in much the same way, one member actively building, largely with moss plucked from neighboring trees, while the other perched nearby, flycatching, calling, and at intervals going to inspect the nest.

On 4 June, I found a completed nest in the same forest where I had watched the flycatcher build on the loop of liana. This nest was attached to the base of a small hart's-tongue fern (*Elaphoglossum* sp.) that grew on a long, dangling, dead liana, which had been severed several feet above the ground. This woody vine hung straight down from a tree at the edge of a *carril,* or avenue cut through the forest to mark the boundary between two properties. The nest was 15 feet above the ground. Above and below it on the supporting liana grew other epiphytes, including ferns, aroids, bromeliads, and much green moss. But these surrounding growths hardly concealed the slight, shallow saucer, which was inconspicuous only because of its small size and because the mosses and liverworts of which the exterior was largely composed blended with those growing on the liana. Mixed with these, in the outer layer of the nest, were some narrow, gray-green, foliaceous lichens. The middle layer was chiefly of fine, dark, fibrous rootlets. The lining consisted principally of the same narrow, branching, foliaceous lichens that entered into the outer wall. The nest measured 3 inches in diameter by 1⅜ inches high. The concavity was 2 inches in diameter by ⅞ inch deep. The high nest built early in the season appeared to be bulkier and more substantial than this late nest, but I could not obtain the former for close examination.

Eggs and Incubation

The nest just described hung above a slope so steep that from its top I could look down and distinguish the eggs. It was impossible to set a ladder on such an abrupt declivity, but by raising a mirror above the nest on a pole I could see the eggs better than from the top of the slope. There were two, dull white, with a conspicuous wreath of brownish blotches around the thicker end.

Three days after I discovered this nest with its complete set of eggs, I started at dawn to watch it, from a point on the slope above it where, as I had proved by a previous trial, my presence did not disturb the incubating parent. She was absent when I arrived at 5:20 a.m., but three minutes later she returned and sat for 19 minutes. Then she left and did not go near the nest for the next hour, when I suspended observations. As I returned at 7:22, the flycatcher entered her nest, and for the next five hours she incubated with the rapid alternation of sessions and recesses typical of a small, exclusively insectivorous flycatcher. On 12 June, the female was again absent from the nest when I arrived at 5:20 a.m., as the dawn seeped into the mist-shrouded forest, dripping from the rain that had fallen through much of the preceding afternoon and night. This time the flycatcher neglected her eggs for more than an hour and a half, or until 6:58, after which she returned and incubated normally until midday. No other flycatcher whose incubation I have studied has acted this way. I can explain her strange behavior only by supposing that the long, hard rains of the preceding afternoon had so reduced her time for eating that next morning she needed to forage continuously for an hour or two to satisfy her hunger.

Only the female incubated. From 7:22 a.m. to 12:22 p.m. on 7 June, she took 43 sessions on the nest, ranging from 0.5 to 12 minutes and averaging 4.4 minutes. Her 42 recesses varied in length from 0.5 to 6 minutes and averaged 2.6 minutes. The most frequent length of both the sessions and the recesses was only 2 minutes. She incubated for 63 per cent of the 5 hours. On 12 June, from her resumption of incubation at 6:58 a.m. until I left at 11:20, the flycatcher took 18 sessions ranging from 0.5 to 19 minutes and averaging 8.2 minutes. The longest session was taken while a passing shower fell, but once earlier in the morning the female sat for 18 minutes while it was not raining. Her 17 recesses varied from 0.5 to 20 minutes and averaged 6.7 minutes. Although the flycatcher was more patient than she had been on the mild morning with intermittent sunshine five days earlier, on this dark, chilly morning she incubated less constantly, keeping her eggs covered for only 55 per cent of the 4 hours and 22 minutes of my watch. If we include in these calculations the 98-minute period of neglect in the early morning, the average length of 18 recesses becomes 11.8 minutes and the flycatcher's constancy for the first six hours of the day falls to 41 per cent—the lowest of any flycatcher that I have studied (Skutch, 1962b, table 2).

To return to her eggs after an outing, the flycatcher alighted on the nest's rim, then without a pause slipped into the bowl and made little sideward movements of her body to adjust the eggs beneath herself. To leave, however, she flew right up and out from the eggs, without first resting on the rim, just as a hummingbird does. While incubating she kept her crest partly lowered but occasionally raised it. She was constantly moving her head around, doubtless looking for flying insects, as Tufted Flycatchers do while perching. Often one that blundered close tempted her to dart from her nest and seize it. Then she might return directly to her eggs, and once during a session of 8 minutes she made three such sallies to catch insects, returning each time. (To avoid complicating the record, I did not count these interruptions lasting only a second or two as "recesses.") Usually, however, if a passing insect lured the female from the nest after she had been incubating for more than a minute or two, she remained away for from one to several minutes to catch more insects.

Twice in the course of the first morning and once on the second morning that I watched, the male gave his mate an insect, each time while she rested on a horizontal vine that crossed the supporting vine below the nest. On one of these occasions, he first took the insect to the nest, in the female's absence, and lowered his head into it, as though anticipating the nestlings (Skutch, 1953:10). Nuptial feeding is not often witnessed in the flycatcher family, but it has been recorded in about eight species (Skutch, 1960:573; Davis *et al.*, 1963: 355).

Except during the long period of neglect at the beginning of the day, both parent Tufted Flycatchers did nearly all their foraging close around the nest, especially in the clear space of the boundary line. They were constantly darting out to capture insects and, as far as I saw, everything they ate was caught in the air. Most of the insects were too small for me to see; none that I noticed seemed more than a half-inch in length. Once the female pecked at a large red bug that was climbing up the mossy supporting vine beside the nest; but after giving a few pecks she paid no more attention to this insect, although it was only a few inches above her while she incubated.

Whenever I raised my mirror above the nest to see the eggs, one or both parents hovered near the intruding object, rapidly repeating slight, sharp notes. Although I never touched the nest, the eggs vanished before they hatched. The incubation period of this flycatcher seems never to have been determined. At earlier nests, however, I had seen that both parents feed the nestlings. In Guatemala, parents attending fledglings out of the nest would sometimes pass an insect to the perching young while flying in front of them, with scarcely a break in their flight—so quick are the movements of these small flycatchers.

SCALY-CRESTED PYGMY-FLYCATCHER

Lophotriccus pileatus

In contrast to many of the smallest flycatchers, which are hard to find and harder to identify, this diminutive, large-headed bird, slightly over three and a half inches in length, makes its presence known by its constant calling and is easily recognized by its voice no less than by its appearance. In the male, the forehead and forepart of the crown are brown. The elongated feathers of the crown, cinnamon-rufous with black centers, form a helmet-like crest which is usually worn flat, as in the Royal Flycatcher. The remaining upper parts of the body are light olive-green. The wings and tail are dusky with olive-green edgings and two conspicuous yellowish bars on the wing coverts. The olive sides of the head are more or less tinged with cinnamon-rufous. The throat and breast are whitish or pale yellow with more or less distinct grayish streaks, and the abdomen is pale yellow. The sides and flanks are tinged with olive-green. The female is similar but has shorter crest feathers. In both sexes, the bill, of moderate length and breadth, is black, with a light tip and base of the lower mandible. The eyes are yellow, and the legs and toes are pale flesh-color.

This tiny flycatcher ranges over the lower slopes of the mountains from Costa Rica to Peru and northwestern Venezuela. In Costa Rica, Carriker (1910:730) found it as low as 1,500 feet, evidently on the Caribbean slope. On the Pacific slope of southern Costa Rica, the lowest point at which I have encountered *Lophotriccus* is 2,500 feet, where it is still quite rare. At 3,000 feet it begins to become abundant, and between 3,500 and 4,000 feet in the Cañas Gordas region I found it common. It is still present in small numbers at 5,000 feet on the Pacific slope, but above this I have no record of its occurrence. Thus in Costa Rica this pygmy-flycatcher is restricted to the humid forests in a rather narrow altitudinal belt. In Venezuela, the species has been found as high as 6,500 feet (Phelps and Phelps, Jr., 1963:218).

The Scaly-crested Pygmy-Flycatcher is a solitary bird which passes most of its life in the lower stories of the woodland, chiefly amid the tall shrubs and smaller trees between about 10 and 30 feet above the ground. Rarely it ascends higher or drops lower, and it may on occasion venture beyond the primary forest into lighter second growth. It subsists largely on insects and spiders, which it espies while perching on a slender twig, then darts up to pluck from a leaf or branch while hovering in the air. After capturing its meal, it does not return to its starting point but alights on a different perch.

Voice

This little flycatcher would seldom be noticed if it did not make such free use of its peculiar voice, which through much of the year is one of the frequent sounds of the mountain forests where it dwells. Most of its notes are sharp, hard, and metallic, loud for so small a bird. Although sometimes given singly, these notes are usually uttered in a series, which may consist of six, eight, or even more of them; the longest sequences are delivered so rapidly that the component notes are difficult to count. At times all the notes of a series are of approximately the same pitch, but more often they ascend, so that the last are conspicuously higher than the first. Frequently the notes are accelerated as they rise in pitch, until the call becomes very rapid and high. A high note may end the performance, or it may be followed by several notes that are lower and more widely spaced. The flycatcher also has a different sort of utterance, a little double whistle which is softer and more melodious, without the metallic timbre of the longer performance.

While calling, the pygmy-flycatcher perches in the lower part of the forest, from about 15 to 30 feet above the ground. He is to be found day after day in the same small area, although he does not restrict his calling to a particular perch. The calling males are rather widely scattered through the forest, even in the regions where they are most abundant; I have noticed no tendency for them to gather in courtship assemblies, as the males of the Oleaginous Pipromorpha and many manakins and hummingbirds do. Yet the persistence with which the Scaly-crested Pygmy-Flycatchers sound their sharp metallic notes through much of the day for months together, and my failure ever to find them in pairs, long ago suggested to me that, like the pipromorpha, the manakins, and the hummingbirds, they are advertising their presence to females to which they will form no lasting attachment.

In 1964 in the Cañas Gordas region, pygmy-flycatchers were already calling rather freely in mid-January. In March and April they were very vocal, but after mid-May I heard them far less often. By early June they called infrequently, and in abbreviated form.

Nesting

Despite much searching through the forests near Cañas Gordas, where the voices of the male pygmy-flycatchers were so often heard,

I saw only one nest. It was in a fairly open part of the woods, near a stream. The pensile structure was attached, 12 feet above the ground, to a slender, leafy twig of a sapling overgrown by the climbing fern *Salpichlaena volubilis.* The nest consisted of a hollow upper part, which held the eggs and nestlings, and a long, thin streamer or "tail" that dangled below the former. The whole structure, from the point of attachment at the top to the tip of the tail, was 15 inches long, but the tail accounted for 10½ inches. The nest proper was only 4½ inches high by 2½ inches wide by 4 inches from front to back, including the visor-like projection that shielded the round doorway in the front. Below the visor, the front-to-back diameter was 2½ inches, the same as the nest's width. The entrance was only 1¼ inches in diameter. The body of the nest was composed of fine, light-colored bast fibers, which in the bottom of the cozy rounded chamber formed a very thick cushion. The body was thinly covered with bright green moss, of which the tail was almost wholly composed. A band of material 2 inches wide securely attached the dangling nest to the supporting twig. In form, this structure resembled the nest of the Sulphur-rumped Myiobius, which, however, is brown rather than green.

When discovered on 9 April, this nest already held two feathered nestlings, resting side by side with their heads at the back of the rounded chamber. When I settled down unconcealed on a log about 20 feet distant to watch, the parent at first complained with a series of staccato notes somewhat like those of the male, but harsher and less metallic. Soon becoming reconciled to my presence, she proceeded to feed her nestlings in silence. All morning, I never had more than one parent, evidently the female, in sight, and only once did I hear distant notes which may have come from a male *Lophotriccus.* From 8:00 to 9:00 a.m., the female fed the nestlings 12 times, and in the next three hours 11, 16, and 18 times, making a total of 57 visits to the nest in the four hours from 8:00 to noon. She delivered all the meals while she clung below the doorway, with her tail pressed against the side of the nest; she did not once enter it. As far as I could see, the nestlings received only insects and spiders, never berries. I recognized orthopterons, moths, and large dragonflies, but other kinds of insects were also brought. I did not see the parent remove droppings; possibly she swallowed them, or possibly I failed to notice them in her bill because her departure was so sudden. In one way or the other, she kept the nest clean. All her movements were swift, and she rarely delayed long in one spot.

A few days later, the nestlings flew away. The presence of only one parent at this nest fits the pattern to be expected from the behavior of the males. It is desirable to confirm this by observations at other nests; but these green, mossy constructions are exceedingly difficult to detect in rainy forests where green moss grows profusely.

MOUNTAIN ELAENIA

Elaenia frantzii

The Mountain Elaenia is a long-tailed, dull-colored flycatcher nearly six inches in length. The sexes are alike, and in both the upper parts are brownish olive. The tail is deep grayish brown. The dusky wings bear two distinct bands of pale yellowish olive and pale olive margins on the remiges and greater coverts. The sides of the head are nearly concolorous with the upper parts, and there is an indistinct pale ring around each eye. The under parts are pale yellowish olive, which merges into pale yellow on the center of the breast, abdomen, and under tail coverts. The bill is yellowish tinged with dusky on the culmen; the eyes, legs, and feet are dark. This flycatcher is confusingly similar in appearance to the Bellicose Elaenia, from which it may be distinguished by its yellowish bill (that of *Elaenia chiriquensis* being black with flesh-color at the base of the upper mandible and on all but the tip of the lower mandible), and more readily by its voice, after one has become familiar with the calls of both species. Fortunately for the bird watcher, the Bellicose Elaenia is a lowland species which scarcely reaches the heights where the Mountain Elaenia dwells. Both of these species differ sufficiently in appearance and voice from the high-crested Yellow-bellied Elaenia, a predominantly lowland bird which overlaps the Mountain Elaenia altitudinally, to prevent confusion by an experienced observer.

The Mountain Elaenia ranges from Guatemala to Colombia and Venezuela. In Guatemala it is rare and known from few localities, such as the ridge between the great volcanoes Agua and Fuego (Griscom, 1932: 273) and the Sierra de las Minas at 6,200 feet (Land, 1962: 276). In Costa Rica, this elaenia occurs chiefly from timberline down to about 4,500 feet in my experience, and occasionally even below 4,000 feet (Carriker, 1910:718). In at least some localities, it is present only seasonally. In 1938, I found it on 18 February on the southern side of the divide between the volcanoes Barba and Poás, at an altitude of 6,200 feet. But on the northern side of the range, in the vicinity of Vara Blanca, I did not become aware of the elaenia until 27 February, although I had been watching birds in this locality since early in the preceding July. Here, between 5,500 and 6,000 feet, it nested, although not so abundantly as higher in the mountains. On the plateau near Cartago at about 4,500 feet, I found the Mountain Elaenia in August and September but not at other seasons. At San Antonio in the Western Andes of Colombia, where another race of this species was common from January to August, Miller (1963:33) could not find it from September to December. Although there is growing evidence that this flycatcher is, at least to some extent, migratory, the details of its movements remain to be clarified.

The Mountain Elaenia lives not only within the heavy highland forests but also, in more conspicuous abundance, in mountain pastures and other clearings with scattered trees and shrubs. Between 6,500 and 7,500 feet at La Giralda, it seemed to be the most numerous bird in the pastures, at least at a distance from the dairy buildings, about which Rufous-collared Sparrows swarmed. Although a number of Mountain Elaenias may gather at a fruiting tree, I have noticed no flock movements. These flycatchers are solitary and unsociable, quarrelsome when they meet; and I have no evidence that they remain mated when not breeding, as is obviously true of the sedentary Yellow-bellied Elaenia.

Food

Although Mountain Elaenias often dart into the air to capture insects in typical flycatcher fashion, perhaps more often they catch their small prey by other methods. They pluck many tiny creatures from the foliage, sometimes entering into the midst of a cluster of leaves and noisily striking against them as they maneuver to seize an insect or a spider. Frequently they cling momentarily to a tree trunk or a stump while they pick something from the bark, or from the moss and lichens that cover it. They also glean insects from slender twigs. Often they alight in a pasture to catch an insect or spider in the grass, then promptly fly up again.

Like other elaenias, this species eats many berries, often swallowing one that seems too large for it to force down. Among the fruits that I have seen it eat are those of *Cestrum, Viburnum, Drimys Winteri, Fuchsia arborescens, Conostegia,* and a small myrtaceous tree known locally as *pisco.* A grove of these elegant trees, abundantly fruiting, seemed to be the chief attraction which drew the elaenias to Las Cóncavas coffee plantation near Cartago in August and September of 1938. They gulped down many of the small red berries.

Soon after their arrival about my residence at Vara Blanca in early March of 1938, the Mountain Elaenias discovered a rich source of food in a tall, fruiting lagartillo tree (*Xanthoxylum* sp.) at the edge of a pasture. The spreading crown was covered all over with broad, flattish clusters of green capsules as large as peas. Each capsule split open longitudinally to expose a single hard seed covered with a thin black aril attractive to a variety of birds. Here came flocks of Mountain Thrushes, uttering sharp *quit quit quit*'s, to perch beside a cluster of fruits and pluck a number of seeds in quick succession. The chief attendants at this feast were flycatchers of several kinds, including yellow-bellied Gray-capped and Vermilion-crowned Flycatchers, nervous little Oleaginous Pipromorphas, and an occasional Sulphur-bellied Flycatcher. But of all the birds that ate the lagartillo seeds, the most numerous and the most constant in their attendance were those little dully clad newcomers to the region, the Mountain Elaenias. They were to be found at the tree at all hours of the day, usually a number together. Sometimes they clung to the low, convex top of a cluster of fruits while they plucked a seed from a newly opened pod, sometimes they fluttered in front of a cluster and snatched a seed without alighting. Unlike the bigger thrushes, they seemed to be satisfied by one seed at a time. They had the same quarrelsome disposition as their relatives the Bellicose Elaenias; and frequently one darted at another, uttering low, angry notes. At times one Mountain Elaenia drove another down into the low bushes beneath the lagartillo tree.

Voice

The note that I have most frequently heard from the Mountain Elaenia is a long-drawn *pee-er,* which is sometimes uttered in a soft, mournful tone, sometimes in a harder, seemingly angry voice. Toward the end of February, I found these elaenias persistently repeating this call in the shade trees scattered through the pastures at La Giralda. One that I watched perched 30 or 40 feet up and called about seven times a minute. The nesting season was approaching, and these birds seemed to be advertising their possession of territory and probably also their need of a mate. As March advanced, these *pee-er* calls became less frequent; and in April and May I scarcely ever heard them except when I visited a nest, or kept the parents from it by my proximity. In these circumstances, they protested with a full, rather deep *pee-er* difficult to distinguish from the note which proclaimed possession of territory in February and March. Apparently these complaining parents were asserting their territorial claims in the face of my intrusion.

In early April, when the *pee-er* became less frequent, I often heard soft, rapid, polysyllabic phrases, a sort of melodious twittering, difficult to describe. These notes seemed to be expressions of contentment or well-being. Sometimes the elaenias voiced low, rattling notes.

I first became aware of dawn-singing in mid-March. In the gray first light of the new day, elaenias perching in scattered pasture trees, sometimes on the topmost twig of a very tall one, called in an odd, dry voice *d'weet, d'weet, d'weet . . . ,* continuing this monotonous chant for many minutes with hardly a pause. Variants of this dawn song were *d'weet, d'weet, d'weet a d'weet . . . ,* and *d'weet, d'weet, d'weeger d'weet. . .* A more divergent rendering was *cheet a cheet, cheet a cheet, cheet a cheet. . .* This bizarre, monotonous dawn-singing continued, on a reduced scale, through most of June. In late May, June, and early July, the type of singing which I had come to associate with the morning twilight was frequently heard long after sunrise, even in bright sunshine, sometimes until the middle of the morning. On 30 June, when the elaenias sang much in the dense cloud-mist which enveloped the mountainside around eight o'clock in the morning, they from time to time introduced into their modest little ditty a low, soft, plaintive trill that seemed to emanate from a melancholy spirit. The nesting season was now waning, and I do not know why the elaenias sang so much in June and early July. Perhaps it was simply an expression of exuberant vitality.

In its dry tone and interminable reiteration, the dawn song of the Mountain Elaenia reminds one greatly of the similar performance of the lowland cousin which it so closely resembles in plumage, the Bellicose Elaenia. The song of the latter is slightly more complex; usually I have paraphrased it as *a we d' de de, a we d' de de*. The Bellicose Elaenia has the harsher voice, and more often gives the impression that it sings or calls in anger. Because of the manifold resemblances of these two species, I confidently looked for the Mountain Elaenia to rise up, singing, in the evening twilight, as I have so often seen the Bellicose Elaenia do. But in spite of the great abundance of these elaenias at La Giralda, I did not once in over four months witness one performing a "sky dance." This is one of the most conspicuous differences between these two sibling species.

Pugnacity

Among the many resemblances between the Mountain Elaenia and the Bellicose Elaenia is their pugnacity. Both are among the most quarrelsome birds that I have studied in Central America. Beginning in March, if not earlier, angry chases, in which two, three, or even four birds participated, were frequent at La Giralda. Often the contestants clutched together and fell to the ground. One morning, while an elaenia rested on the grass in a pasture, another repeatedly pounced down on it from a height of about a foot. The bird in the grass held up its open bill in threat, and appeared to suffer no injury. On an afternoon in late May, I saw four elaenias fall in a cluster from a shade tree to the pasture grass beneath it. Then, while two remained here close together, a third pounced down on one or both of them from a height of a few inches, repeating this several times. Meanwhile the fourth elaenia rested on the grass a few inches from the victims of this assault. In less than a minute, all flew away.

The belligerence of the Mountain Elaenias, their hot pursuits and frequent clashes, fit into the picture of a migratory or at least a wandering bird which claims territory and acquires a mate as the nesting season approaches. Constantly mated birds which have long been settled on their territory tend to arrange matters with their neighbors without such conspicuous conflicts. The pugnacity of the Bellicose Elaenia is also associated with a wandering habit. The sedentary, constantly mated Yellow-bellied Elaenia impresses one as a far milder, more pacific bird.

Nest

At La Giralda in 1963, building began early in April. Nests were placed in a wide variety of situations. Many were built in trees standing in pastures, either isolated or with a few others. These nests might be high or low, on horizontal or upright limbs, excellently concealed by foliage or in exposed situations, rarely on dying trees with little foliage to screen them. Of the 18 nests that I saw, five were placed in the tall, cane-like bamboos so abundant in the forests at high altitudes. These nests in bamboos were usually at or near the woodland's edge; they were situated either near the end of a slender, gracefully drooping leafy spray or amid the lower, erect portion of the canes. In height, the 18 nests ranged from 6 to about 50 feet, with an average of about 20 feet. One of the lowest nests was built among vines (*Muehlenbeckia tamnifolia*) which thickly covered a high stump in a pasture. The highest nest was on a short, thick, moss-covered stub of a branch at the top of a dead tree. Here it was supported between the slender stems of an epiphytic shrub, and shaded by a large tank bromeliad.

The nests of the Mountain Elaenia are small, compact, open cups. The exterior is usually well covered with green moss or liverworts, sometimes with a liberal admixture of finely branched beard lichens (*Usnea*) or of greenish gray foliaceous lichens. In some nests, the green envelope is sparse and other components of the outer shell, such as dark fibrous rootlets or light-colored, decaying, fibrous leaf-sheaths, add their color to the outer surface. Here one sometimes finds wefts of cocoon silk of various hues, gray, yellow, or brown, and perhaps a few small downy feathers. The inner layers and lining are usually composed of many dark fibrous rootlets, or the shiny black fungal rhizomorphs known as "vegetable horsehair," or of both kinds of filaments together. Mixed with these strands are often a few downy feathers. Lying loose in the bottom of the nest are usually some downy feathers—rarely as many as six—some or all of which may be deposited there after the eggs were laid. Nests range in size from 3½ to 4 inches in over-all diameter, 2 to 2½ inches in height, 2 to 2¼ inches inside diameter, and 1 to 1½ inches in depth. Even in a single locality, there is considerable variation in the bulk and composition of nests. Some are rather massive, and rarely one is so slight that the eggs can be glimpsed through the bottom. Some nests of the Mountain Elaenia can be closely matched by nests of the Bellicose Elaenia; but it is hardly possible to confuse the nest of either of these species with the neater, lichen-covered cups, with a generous lining of downy feathers, built by the Yellow-bellied Elaenia.

I was unable to make a thorough study of nest construction, but at a late nest where building proceeded slowly, I saw no indication that both parents shared the task. Evidently the female builds alone, but the male sometimes comes to examine her work.

Eggs

Of 12 nests whose contents could be seen, 11 contained two eggs or nestlings. In one nest that was kept under observation throughout the period of laying, only one egg was ever seen. In another nest, the first egg was laid between 8:00 and 11:15 a.m. on 16 May, and the second was laid between 7:40 and 10:20 a.m. on 18 May. At a different nest, the second egg was laid between 6:45 and 9:00 a.m. Flycatchers often lay their eggs later in the morning than finches and tanagers do. The eggs are dull white to pale buff in ground color, marked with spots and blotches of pale cinnamon, rusty brown, or chocolate, which usually form a wreath around the broad end and are thinly scattered elsewhere. In shape the eggs may be short ovate and blunt or more elongate and pointed. The measurements of seven eggs average 20.3 by 15.7 mm. Those showing the four extremes measured 22.0 by 15.0, 19.9 by 16.1, and 19.1 by 15.6 mm.

In 17 nests in the Cordillera Central of Costa Rica, 5,500 to 7,500 feet above sea level, eggs were laid as follows: April, 9; May, 6; June, 2.

Incubation

At a nest which I watched from 5:53 a.m. to 1:05 p.m. on 9 May 1963, I saw no evidence that more than one bird took an interest in the eggs. In this species, apparently only the female incubates, as in all other flycatchers that have been carefully studied (Skutch, 1960). The early morning was sunny, with a cold wind blowing across the mountain much of the time. After the middle of the forenoon clouds gathered overhead. By 10:40 a fine drizzle was falling, and from 11:30 until I left there were intermittent showers, sometimes hard. Despite these changes in the weather, the elaenia preserved a fairly steady rhythm of incubation. In slightly over seven hours, she took 14 sessions, ranging from 9 to 62 minutes in length and averaging 25.1 minutes. Her longest session was taken from 7:23 to 8:25, while the sun shone and a cold wind blew; her next longest session was only 36 minutes. Her 14 recesses ranged from one to 9 minutes and averaged 5.2 minutes. After rain began, the female left her eggs when the showers stopped and returned when they were renewed. And so, through rain and shine, the little elaenia kept her eggs covered for 83 per cent of the seven hours that I watched her, which is exceptionally high constancy for a flycatcher. As she approached her nest through the dense foliage of the thorny tree that held it, she usually voiced a few whistled *pee-er*'s.

At two nests, the incubation period was 15 days.

The Nestlings

The newly hatched nestling has skin which is yellow to pinkish orange in color, and it bears sparse, dark gray down of the usual passerine type. The nestling's bill, tarsi, and toes are bright yellow, and its eyes are tightly closed. When the nestling is between three and four days old, the sheaths of its primary feathers and rectrices begin to push through the skin. Two days later, pinfeathers begin to emerge from the body as well as the wings and tail. The eyes are still closed. At the age of eight days, the feathers begin to expand from the ends of their sheaths nearly everywhere, except on the head. The eyes can at this age be partly opened, but they seem to be kept closed most of the time. At ten days, the young elaenia, open-eyed and alert, is fairly well clothed with plumage. To my surprise, the nestlings that I examined developed no secondary or interpterylar down. Such down is a prominent feature of week-old Yellow-bellied Elaenias and is more sparsely developed on nestling Bellicose Elaenias (Skutch, 1960:303-304, 316); yet these birds of lower and warmer regions seem to have less need for this additional covering than do the Mountain Elaenias.

The young are fed by both parents. A nestling which grew up alone in a low nest, hidden in a clump of bamboos, left when only 14 days old. But two that were raised in a higher and more exposed situation remained considerably longer. The first left when about 17 days old. When I raised a mirror above the nest to see whether its occupants were still present, the second sallied forth and flew about 40 feet, to alight on the ground. Here I caught it for a closer view, and after I had finished my examination I placed it in a tree, whence it promptly took flight and covered about 100 feet on a descending course, again landing on the pasture grass.

When they leave the nest, the young elaenias resemble their parents in plumage but are browner above. The interior of the mouth is orange-yellow, and the corners yellow. The eyes are black. The legs and toes, which were bright yellow when the nestling hatched, have become dusky, and the claws have blackened.

SLATY-CAPPED FLYCATCHER

Leptopogon superciliaris

A slender, dull-colored inhabitant of the forest, the Slaty-capped Flycatcher is about five inches long. In both sexes, the whole top of the head is dark slate-color. The remaining upper parts are largely olive-green, with a grayish brown tail. The dusky wings bear two rows of yellowish spots on the coverts, forming broken bars, and the remiges have conspicuous yellow margins. The cheeks are pale gray with darker speckles, and there is a conspicuous blackish patch behind the auricular region, separated from the slaty occiput by a narrow band of gray. The throat is olive-gray, passing into yellowish olive on the chest and sides and into purer yellow on the abdomen. The short, narrow bill is black with a light tip, and the eyes are brown. The slaty pileum and the black patch on the auricular region are good recognition marks of this flycatcher whose olive-green and yellowish plumage resembles that of many other species.

This flycatcher ranges from Costa Rica to Bolivia, northern Brazil, and Trinidad. On the Pacific slope of Costa Rica I have rarely seen it as low as 2,500 feet, but between 4,000 and 5,000 feet it is moderately abundant. I have no precise information as to how much higher than this it extends, nor about its altitudinal range on the Caribbean side of the country; during my year at Montaña Azul, I failed to find the bird as high as 5,500 feet. In Venezuela, different races occur from the Tropical Zone upward to about 6,250 feet (Phelps and Phelps, Jr., 1963; 243-244).

I have found the Slaty-capped Flycatcher in the forest and at its edges, never in open country. Usually I have seen it perching well above my head, from about 15 to 30 feet above the ground; but it is a restless bird and has rarely delayed in one spot, or even remained in view long enough for me to distinguish the details of its plumage. It has always been the only individual of its kind in sight, but often it has been a member of a mixed flock of small forest birds, such as the Slaty Antwren, Tawny-crowned Greenlet, Gray-headed Greenlet, Red Ant-Tanager, Sulphur-rumped Myiobius, and Spotted Barbtail. From a slender, rather exposed branch, the Slaty-capped Flycatcher makes swift darts to snatch insects from the air or the surrounding foliage.

This flycatcher's call, which it gives sparingly, is a disyllable that sounds like *peet-yer,* uttered in a most peculiar tone, as though the elusive little bird were making disparaging remarks about its companions in the mixed flock, or perhaps about the watcher who tries ineffectually to keep it in view. Other notes are *hit chú, hit chú,* like a little sneeze; *hit cheee,* with the second syllable thin and long drawn out; and sharp little monosyllables. Slud (1964:274) heard this flycatcher deliver trilled or rippled notes.

Nesting

The Slaty-capped Flycatcher is another of those retiring forest birds whose rarely-found nests make a more lasting impression than an encounter with the birds themselves. The single completed nest of this flycatcher that I have seen was situated in a wildly picturesque setting. From my cabin near the confluence of the Pacuar and San Antonio rivers at the head of the Térraba Valley, we had worked far upward along the San Antonio and then its tributary that I came to call "The Hummingbirds' Brook," from the number of nests of the Violet-headed Hummingbird that we found along its course (Skutch, 1946, 1958a). Toiling slowly up into the coastal hills along the narrow rocky bed of the brook, through high unbroken forest, we came after several hours to a point where a vertical wall of rock, about 10 feet high, rose ahead in the middle of the channel, dividing the limpid current into two separate streams. That on the right fell with a single leap into a deep, sandy recess in the rock. The left branch babbled down through great loose boulders, beneath a huge block which, wedged between the central pier of stone and the high, rocky wall of the ravine, formed a bridge over the cascade. Like most of the larger rocks above reach of the flood waters of this forest stream, this block was profusely overgrown with ferns, begonias, aroids, clusias, and other air plants. To reach the top of the wall that obstructed the channel, we were obliged to scramble beneath this great angular block of stone. Just upstream from it, a huge, shattered trunk lay in the narrow streambed, between sheer rocky cliffs. The base of the trunk projected over the rapids that passed beneath the bridging rock. From a stout splinter on the lower side of the trunk hung a nest that we should never have noticed, if we had not been obliged to clamber up the rapids beneath the natural bridge.

The strange nest hung free, 4 feet above the boulders among which the cascade leapt. It was between globular and pyriform in shape, with a round entrance in the side, shielded by a visor-like projection that seemed superfluous in this sheltered situation. Including the elongated point of attachment at the top, the nest measured 8 inches in length. From side to side it was 3½ inches thick, and from the back to the edge of the projection over the doorway it measured 6 inches. The thick, dark-brown walls were composed almost wholly of fibrous rootlets, with a small admixture of light-colored fibers. The ample globular egg chamber was lined all around with light-colored bast fibers finely shredded and some tufts of silky seed down.

Within this nest lay, on 13 March 1940, a single nestling in pinfeathers. The long gray tufts of down that terminated some of these pins had evidently been present when the nestling hatched. The interior of its mouth was orange-yellow. It cried shrilly when I looked in at it with a lighted electric bulb and a small mirror.

The most distant point whence I could satisfactorily watch this nest was only about three yards away. After I had sat here, necessarily unconcealed, for many minutes, a parent fed the nestling. Although very active and nervous, it was not shy. It brought food only thrice in three hours, but these were the warmest, drowsiest hours of the day. The three insects that were taken to the nest were all green, leaf-dwelling kinds—orthopterons that resembled katydids and tree crickets. Their substantial size explained the infrequency of the feedings. When approaching and leaving the nest, the little olive parent darted beneath the epiphyte-covered bridging rock, thereby making its sudden movements still less conspicuous to whatever hostile eyes might be lurking amid the vegetation overhead or on the surrounding rocks. There was never more than one parent in sight; and I should not be surprised if in *Leptopogon,* as in a number of other small flycatchers, only the female attends the nest. However, the difficulty of reaching this nest along miles of rocky streambed discouraged the long-continued watching needed to prove this point.

On Trinidad, Belcher and Smooker (1937:248-249) found a good many nests of the Slaty-capped Flycatcher. In the Northern Range of this island, the flycatcher builds "in a dark place such as a deep cleft in a huge riverside boulder, or the little caves where earth has fallen away just under the top of a high bank at the side of a road and a fringe of vegetation hangs down the front like a screen." In such situations the nest, in shape evidently much like the one I found in Costa Rica, is suspended from a short length of root or tendril. It is "built of all manner of vegetable fibers well felted together, with always a certain amount of parti-coloured cocoons on the outside, by which it can at once be known, and a lining of fine soft materials in the egg-chamber." Eggs were found from early February to early July. Never more than two in a set, these eggs were lighter in weight and thinner in the shell than those of any other flycatcher of approximately equal size known to these authors. They were unspotted white, and the measurements of six averaged 19.5 by 14.1 mm.

Nests of the related Sepia-capped Flycatcher found in the state of Veracruz, Mexico, by Moore (1944) also hung beneath huge rocks or logs in dark situations. In shape, they resembled those of the Slaty-capped Flycatcher, and each was suspended from its overhead support by two parallel roots or fibers—a rather unusual mode of attaching a flycatcher's nest. One nest contained three pure white eggs and another held two feathered nestlings.

Family CORVIDAE

STELLER'S JAY

Cyanocitta stelleri

This beautiful jay, slightly over eleven inches in length, is nearly everywhere deep blue. Above each eye is a broad white line, which does not extend to the base of the bill, and the lower eyelid is also white. Its cheeks are black and its chin and throat are grayish white. Its long blue crest can be raised high but is usually carried flat. The bill, legs, and feet are black and the eyes are brown. As in most jays, the sexes are indistinguishable. This description applies to the race *C. stelleri ridgwayi* of Guatemala, which is a much bluer, more handsome bird than the Steller's Jays familiar to residents of the western parts of temperate North America.

Steller's Jay ranges from southern Alaska and the Rocky Mountains in the United States south to Nicaragua. Since so much has been written about the northern races of this jay but so little about the southern ones, I shall confine this account to what I learned about the habits of the Guatemalan race during the 13 months when I saw and heard it daily on the Sierra de Tecpam and also on later journeys through the Guatemalan highlands. In this country it is common in open stands of oaks and pines and cultivated areas with scattered trees, from about 5,000 to 11,000 feet above sea level. Below 5,000 feet it is replaced by the Bushy-crested Jay as the common representative of the family outside the heavy forest. In October of 1933, I found Bushy-crested Jays abundant among the shaded coffee plantations about the shores of Lake Atitlán, 5,000 feet above sea level. Here Steller's Jays were rare; but as one travelled up the valley of the Río Panajachel that flowed into the lake, they became increasingly numerous, and on the steep slopes above the valley they were the only jays I saw. On the other hand, the elegant Bushy-crested Jays, which preferred the more luxuriant subtropical growth near the lake, rapidly disappeared as one climbed to the uplands. But on a scrubby mountainside above Nebaj on the wetter northern slopes of the Sierra Cuchumatanes, I found these two species of jays flocking together at 7,800 feet—the highest point that I have seen the Bushy-crested Jay. Among the heavy cypress forests near the summit of the Sierra de Tecpam, 9,000 to 10,000 feet above sea level, Steller's Jays were scarce and lived chiefly about houses; but on the Sierra Cuchumatanes they were not uncommon, in the open stands of pine far from human habitations, up to at least 11,000 feet.

Strangely enough, in northern Nicaragua Steller's Jay occurs down to 3,500 feet, much lower than it has been recorded in Guatemala

(Griscom, 1932:403). Farther north, the species ranges upward to 13,000 feet on Mount Popocatépetl in Mexico (Miller *et al.*, 1957:127) and, at the other extreme, it descends to the shores of the Pacific from central California to southern Alaska.

In the pretty little garden beside the cottage that I occupied on the Sierra de Tecpam, as about the surrounding pastures and cultivated fields and in the open stands of pines, oaks and other trees, Steller's Jays were abundant. Indeed, their numbers, large size, bright plumage, active habits, and raucous voices that were seldom silent, made them the most conspicuous birds in the neighborhood. All day long, but especially in the early morning, they were to be seen flying from tree to tree in loose, straggling flocks. Although sociable birds, perpetually chattering to each other, each preserved a certain independence of movement, like toucans; they were individualists who disdained to fly in compact flocks, as pigeons and parrots do. Most of the time their blue crests were laid back inconspicuously, to be raised as an expression of surprise, curiosity, anger, or some kindred emotion.

Voice

Steller's Jays and Great-tailed Grackles are about the two noisiest birds in Guatemala, and when both are present in large numbers, as they were on the plains around Tecpam some 7,000 feet above sea level, the woods and fields are rarely silent. How annoying these loud-mouthed birds can be, when one wishes to enjoy the softer, sweeter voices of other birds! One expects to hear, from a bird so chastely splendid in attire as the Guatemalan Steller's Jay, notes more genteel than the rasping *chaaa chaaa*—like the sound of a file rubbing against the edge of a thin metal plate—that it repeats all day long. This call gives the jay its Guatemalan name, *Charra*. Another frequent utterance is a series of grating nasal mews that reminded me of the Catbird's call or the autumnal complaints of a lone gray squirrel.

Occasionally Steller's Jay reveals that it can utter softer sounds, one of which resembles the *willica* of the Blue Jay, but is neither so loud nor so ringing. One afternoon I found a Steller's Jay perching in an oak tree, attempting to sing in a mellow voice and almost succeeding. Possibly it was trying to reproduce the clear whistles of the Yellow-backed Oriole—a feat too difficult for it. As I sat writing in the cottage around noon on several days in early March, a jay rested in the shade outside, singing to itself in an undertone. Its medley of low, soft notes, punctuated at intervals by a little rattle very characteristic of its kind, was such a sweet, pleasing performance that I wished to hear every syllable of it; but the inconsiderate companions of this songster were all the while chattering so loudly and harshly that its subdued voice was scarcely audible to me. Similar whisper songs have been noticed in a number of corvids that never sing loudly. These birds seem to sing solely for their own comfort, as a man hums a tune while he goes

FIG. 3. Panoramic view on the Sierra de Tecpam, Guatemala, at about 8,500 feet above sea level. Steller's Jays roamed through the pines, oaks and other broad-leafed trees and over the adjoining pastures and cornfields. In the moister parts of the woodland lived Black Thrushes, Brown-backed Solitaires, Chestnut-capped Brush-Finches, and other birds of the altitudinal temperate zone. 10 December 1933.

about his work, and stops short if another person comes within hearing.

FOOD

To fill their stomachs, Steller's Jays let no opportunity slip by. Few birds are more versatile foragers. They often hunt over the ground, scratching in bare places such as roadways or rummaging among fallen leaves in the woods. They probe for insects and other small creatures among the lichens and mosses which form a thick mantle over many of the boughs of the oak trees. They eat berries of various kinds and are especially fond of acorns. Unlike its neighbor the Band-tailed Pigeon, which swallows these fruits whole, the jay opens them to extract the meat. Carrying an acorn to a thick branch, the jay holds it down with a foot while it pecks at the shell, throwing the weight of its whole stiffened body into the hammer-like movement. Acorns not immediately wanted may be hidden away for future use. One foggy morning, I watched a jay place an acorn in the angle between a branch and the trunk of a pine tree, about 12 feet above the ground, then cover its cache with pine needles and lichens. When I climbed into the tree, I could find no other acorns hidden there. I heard many complaints about the "Charra's" depredations on the ripening maize; but in a large cornfield that I kept under observation, I noticed little damage. The jays preferred the acorns which at the same time were ripening in the surrounding woods.

Nesting

Despite the jay's abundance on the Sierra de Tecpam, I found few of their well-hidden nests. The first of these was 20 feet up in the top of a young pine tree growing at the edge of a bushy clearing, beside open woods of oak and pine. This nest was a bulky cup, composed of coarse twigs, coarse uprooted grass plants, lichens, green moss, and one twig with attached dead leaves that apparently had been brought while still green. The interior of this nest, which measured 4 inches in diameter by 2½ inches in depth, was well lined with fine fibrous roots and grass stems. Here, on 30 April 1933, lay two, beautiful, light blue eggs, heavily blotched with pale lilac and olive, especially on the thicker end. They measured 33.3 by 22.6 and 32.5 by 23.0 mm. Eager to study the behavior of the parents at this nest before the eggs hatched, I removed some of the pine needles which screened it too well for my purpose. Unhappily, the jays deserted their eggs, which I then opened, to learn that they had been freshly laid.

My second nest was in a similar situation, 20 feet up in the top of a young pine too slender to climb, in open pine woods. On 7 May it held two nestlings, already well feathered, which I could glimpse from the top of a neighboring tree. Returning after nightfall, I again climbed the neighboring tree, and in the beam of my flashlight saw a parent covering the nestlings, who so completely filled the nest that she was above rather than in it. The gleam of an eye revealed that she was awake and alert, but she bravely stuck to her post. Three nights later, these young jays, now ready to fly, were still being protected by a parent from the nocturnal chill of these high mountains. In their juvenal plumage, the young rather closely resembled their parents.

On 12 April 1958, Baepler (1962:147) found two nests of Steller's Jay in oak-pine forest near Soloma in the Department of Huehuetenango, Guatemala. One, at an altitude of 8,400 feet, was about 20 feet up in an oak tree and contained two eggs. The other nest, at 6,400 feet, was 15 feet above the ground and had only one egg. These eggs are described as bluish, slightly speckled with brown.

BLACK-THROATED JAY

Cyanolyca pumilo

This lovely jay, about ten inches long, is clad almost everywhere in deep, rich ultramarine. A patch of black covers its forehead, the sides of its head, and its throat. Separating the black of the forehead and cheeks from the blue of the crown is a narrow line of white, which extends backward above the eyes and fades away over the ears. There is no crest. Its bill, legs, and feet are black. The sexes are alike, as is usual in jays.

The Black-throated Jay ranges from the Mexican state of Chiapas through Guatemala to Honduras and El Salvador, and in altitude from about 6,000 to 10,000 feet above sea level. On the Sierra de Tecpam, where alone I have seen this small jay, it was strictly confined to the woodland and never ventured forth into clearings with scattered trees, where Steller's Jays were so numerous. Although sometimes I found Black-throated Jays in open stands of oaks, pines, and alders, they preferred the heavier, more humid forests, such as were to be found in sheltered ravines, where the fine old trees, largely oaks and pines, were veritable gardens of epiphytes, including bromeliads, orchids, ferns, mosses, lichens, and many other kinds of aerial growths.

In such ancient, unspoiled forest, I found Black-throated Jays in flocks consisting of from half a dozen to a dozen or more birds, and on several occasions was able to watch at close range their interesting methods of foraging, as they were less excitable and suspicious than many other jays. Sometimes for a long while they confined their searching to the foliage of the lower boughs of the trees, the undergrowth, and the vine tangles. They moved slowly among the leaves clustered at the ends of the branches, carefully scrutinizing them for adult and larval insects. At times one of the foraging jays was completely hidden in a dense mass of foliage. Often the bird clung to the very tip of a thin twig while it examined the surfaces of the leaves. These jays also stooped over to investigate the sides and bottoms of the slender branches along which they hopped—a mode of foraging widespread among the lovely little painted tanagers of the genus *Tangara*. When a jay found a curled leaf, living or dead, he plucked it and held it against a branch with one foot while he pulled it apart with his bill to see what it contained. Often his search was rewarded by the discovery of a larva or a spider that had taken sanctuary in the curled leaf. The jays investigated the foliage of the epiphytes as well as that of the trees that supported these growths. At times one ascended an upright branch like a creeper, probing among the lichens that covered it. The search for insects and spiders was thorough and methodical, neglecting hardly any part of the vegetation, above the lowest stratum, where such small creatures were likely to be lurking; although often the jays reminded me of overgrown blue vireos as they hunted through the foliage, their foraging was more versatile than that of such small birds as vireos and wood warblers.

While hunting, the jays talked quietly together in low, whining notes, interspersed with sharper, brisker calls, much like those of the Unicolored Jay, but not so loud. These woodland jays seemed far more gentle and well-bred than their boisterous, garrulous cousins of the neighboring clearings, the irrepressible Steller's Jays. Even their alarm note, although typically jay-like, was quieter than that of many members of their tribe. To my regret, I discovered no nest of this charming jay, and have found nothing about its breeding in print.

Family TURDIDAE

MOUNTAIN THRUSH

Turdus plebejus

This retiring inhabitant of the mountain forests is about nine inches in length and plainly clad in shades of brown. In both sexes, the brown is deepest on the head, where it is nearly sepia, olivaceous on the remaining upper plumage, and paler on the under parts. The throat is indistinctly streaked with deeper brown; and each grayish brown feather of the under tail coverts has a central wedge-shaped area and broad margins of pale brownish buff. The bill is black; the eyes brown; the legs and feet dark. This thrush can best be distinguished from the confusingly similar Pale-vented Thrush by its spotted or scaly under tail coverts. In the latter, the abdomen and central under tail coverts are plain white and contrast conspicuously with the deep brown of the flanks. The Pale-vented Thrush is usually found at lower altitudes than the Mountain Thrush, but their vertical ranges may meet.

The Mountain Thrush ranges through the highlands from the southern Mexican state of Chiapas to western Panama. In Guatemala, where I never knowingly saw this bird, it has been recorded chiefly in the cloud forests between 6,000 and 9,000 feet. In Costa Rica, where I have found this thrush from about 4,500 to 9,000 feet, it is abundant in the wilder parts of the highlands. Here it lives in forests of towering trees heavily burdened with mosses, liverworts, ferns, and flowering epiphytes, especially along the woodland edges, whence it ventures forth into pastures and other clearings with scattered, lichen-encrusted trees and epiphyte-covered stumps, to forage and even to nest. It may even visit open stands of low trees far from the heavy mountain forests, if these provide some attractive fruit. In August and September of 1938, I found Mountain Thrushes on the plateau between Cartago and Paraiso, amid extensive pastures and plantations, in a grove of the handsome myrtaceous tree known locally as *pisco,* whose small red berries were sought by a variety of birds. At other seasons, I failed to meet the Mountain Thrush in this district.

In 1963, I found Mountain Thrushes numerous on the La Giralda dairy farm high on the western end of the massif of Volcán Barba. On April mornings, I would see thrush after thrush fly swiftly down the long grassy slopes, between scattered trees, toward the Central Plateau far below. Often they flew near the ground, sometimes passing between the trunks of the trees along the forest's edge. I never saw these thrushes in flocks, but each traveled alone, continuing downward in a direct and purposeful manner until lost to view. This movement continued until past the middle of the morning, when already other Mountain

Thrushes were returning up the slopes. The upward passage, however, was most conspicuous in the afternoons. Evidently some fruiting tree or shrub on the lower slopes attracted the thrushes, but I did not succeed in following them far enough downward to learn what it was.

In March of an earlier year, however, I had watched Mountain Thrushes come in numbers to a fruiting lagartillo (*Xanthoxylum*) tree growing at the forest's edge at 5,500 feet on the northern slope of the same range. Perching at the top of the tall tree beside a broad, flattish cluster of the green pods, now splitting longitudinally to expose little bony seeds enclosed in a thin black aril, a thrush would nervously pluck and swallow a number of the seeds in swift succession. In addition to a variety of fruits and arillate seeds, these thrushes eat insects and other small invertebrates, which they sometimes gather while hopping over the pasture grass close by woodland, especially when clouds drift close to the ground, making the birds less conspicuous as they hunt at a distance from cover.

Voice

The sharp, nervous *quit quit* or *whip whip* of the Mountain Thrush, often uttered in flight, reminded me strongly of the notes of the American Robin and the Rufous-collared Thrush of northern Central America. A parent Mountain Thrush flying from its nest called *whip whip whip whip wic wic,* an accelerated sequence of sharp notes. Alarmed for the safety of their young, parents complain endlessly with *tock tock tock,* interspersed with occasional *whip whip*'s. In a questioning or uncertain mood, they voice a low *toc;* and they may combine the *tock* and *whip* notes in various patterns.

There is a certain aristocratic elegance in the Mountain Thrush's plain, rich brown plumage, which seems in keeping with the character of a bird who spends much of its life amid the gray cloud-mist which periodically invades its moss-draped forests. But its song, the least melodious that I have heard from any member of the thrush family, is remarkable for nothing except monotony.

About the beginning of April, 1938, while I dwelt amid the cool subtropical forests at Montaña Azul, I began to hear bird notes which puzzled me greatly. The tireless reiteration of a few unmelodious notes suggested a hummingbird, but the voice was somewhat stronger than that of most members of this family. Even when, through the gray clouds, I at last glimpsed the author of these baffling notes, noting his size and form although his colors were dimmed by the mist, I never suspected that I saw and heard a member of a family famed for its vocal accomplishments. Only on a later and more favorable encounter with the monotonous performer was the surprising truth forced upon me.

The song of the Mountain Thrush is a rapid, long-continued—almost an endless—succession of clear but weak and characterless notes, with little variation in pitch and the merest suggestion of rhythm. *Chir chip chip cher chip chip cher cher chip chip* . . . he sings; it would be tedious to spell out all the notes of this sort which compose a song. While singing, the thrush usually perches high in a tree, where he remains motionless and is difficult to detect. If he discovers that he is seen by human eyes, he suddenly ceases, utters a few sharp, robin-like notes, and flies away. He sings chiefly at dawn and in the early morning, and on dim, misty afternoons. On the dull afternoons so frequent in these mountains in April and May, the artless songs of the Mountain Thrushes float out of the cloud-mist on all sides, with a monotonous persistency that hummingbirds can hardly surpass. But in these months which the majority of the birds find most favorable for nesting, there are days of almost continuous wind-blown rain which silence even this poor songster who loves to raise his voice in sad gray weather.

At La Giralda in 1963, as at Montaña Azul in 1938, the Mountain Thrushes began to sing in early April and fell silent in the first week of June, after little more than two months of song. By late June they were molting heavily.

Nesting

The Mountain Thrush makes such good use of the excellent opportunities that the epiphyte-laden trees of the mountain forests offer for concealment, that in two seasons passed in localities where these thrushes were abundant I found only two of their nests, both after the young had hatched. The first of these nests, discovered at an altitude of 5,600 feet on 25 May 1938, was about 40 feet above the ground, in the midst of a great, intricately entangled mass of vines and large epiphytes, all overgrown with moss, that hung from the trunk of a tall tree standing amid low, bushy growth. The nest itself was both invisible and inaccessible to me, but I was sure of its presence because the parents repeatedly entered the tangled mass of vegetation with food.

I passed the whole day of 5 May 1963 watching a nest of the Long-tailed Silky-Flycatcher in a pasture beside the oak forest, at an altitude of 7,500 feet. From early morning a pair of Mountain Thrushes frequented the vicinity, often with overflowing bills of food; but it was long before I could discover where they were taking it. Whenever my eyes were turned toward these parents, they hung around with full bills, complaining; but while my attention was directed to the silky-flycatchers, or I made an entry in my notebook, the food would disappear, evidently having been delivered to nestlings. Thus the whole morning passed without my even discovering in which tree their nest was situated. In the afternoon, however, I learned that the parent thrushes, or one of them, had developed a clever way of visiting the nestlings without betraying their location, in a blasted but still living, epiphyte-burdened tree only 20 feet from where I sat. The parent would fly into an epiphytic shrub well above the nest, then work its way unobtrusively down through the dense foliage, whence I heard its movements rather than saw the bird. After it had been out of sight a good while, it would fly from the tree so suddenly that I was uncertain just where it came from.

I did not locate the nest until the following morning, when I climbed into the tree and with some difficulty distinguished it in a dimly lighted crotch, 10 feet above the ground. In this broad niche between thick branches it was darkly shaded and excellently concealed by epiphytes and by two tall, densely foliaged water sprouts that grew in front of it. The roomy cup, embedded in the dead leaves and other debris that had lodged in the crotch, was composed externally of green moss. Next within was a layer of dry, narrow, light-colored bamboo leaves, such as were available at the forest's edge 50 feet away. Within this were a few black rootlets or similar strands, not enough to cover the light-colored leaves. The interior of the cup was $3\frac{1}{2}$ inches in diameter by $2\frac{7}{8}$ inches in depth.

Here lay two fat nestlings with pinfeathers just sprouting. The inside of their mouths was orange. The pale yellow flanges at the corners were remarkably long and broad, extending from behind the eyes to the nostrils, and measuring about 14 mm. in length by 5 mm. in breadth at the base of the bill. While I examined the young thrushes, their parents complained with varied notes, as earlier described.

When I again watched the silky-flycatchers' nest on the morning of 19 May, the parent thrushes as usual protested my presence so close to them but apparently fed their nestlings. Suddenly, at 8:20 a.m., a young thrush flew out of the nest tree to alight on a prostrate log in the pasture. After a while it dropped to the ground, rested there for over an hour, then hopped up to some sprouts on the side of a trunk. From time to time a parent brought it food. Here I visited it after I finished watching the silky-flycatchers. The fledgling permitted me to approach within arm's length, then, uttering a single, slight, sharp note, it suddenly took wing and flew easily about 100 feet down the slope, to alight in a large arborescent fuchsia. It had remained in the nest 13 days after I found it with sprouting pinfeathers. Now its plumage closely resembled that of its parents, except for some rather inconspicuous buffy spots or bars on its breast and wing coverts. Four or five excited silky-flycatchers gathered around the young thrush, but they did not go close to it.

At the end of May, on the slopes of the Barba massif between 7,000 and 8,000 feet, Mountain Thrushes were everywhere feeding fledglings, from nests for which I had searched in vain.

BLACK THRUSH

Turdus infuscatus

Although absent from temperate North America, black or blackish thrushes, some of which resemble the European Blackbird, are widespread in the Andes of South America. In Costa Rica, the large Sooty Thrush is a familiar sight on the higher summits, above 8,000 or 9,000 feet. The northernmost of these dusky thrushes is the Black Thrush, which ranges from El Salvador and Honduras to the mountains of eastern Mexico. About nine inches long, the male is wholly black in plumage, with the bill, eyes, legs, and toes bright yellow. The female is olive-brown, paler on the under parts, with a dull whitish throat streaked with dusky brown. Her legs and feet are yellow as in the male, but her bill is dark. This and related species are sometimes called "robins" or "ouzels," but it seems proper to retain "thrush" as the general designation for members of the nominate genus of the thrush family.

In Mexico, the Black Thrush has been found at altitudes ranging from about 2,000 to 11,500 feet (Miller *et al.,* 1957:185-186). In Guatemala, it ranges at least from 4,000 to 10,00 feet above sea level. In the Guatemalan highlands, where the Rufous-collared Thrush is the familiar "robin" in pastures and cultivated fields, the Black Thrush lives in the mountain forests, shuns the open country, and is accordingly little-known. It is evidently migratory, for on the Sierra de Tecpam I noticed its presence only during the breeding season. I saw my first Black Thrush at the beginning of February, high in a moss-draped tree in the cloud forest. In March, April, and May, the loud songs of these thrushes rang through all the woodland; but in June they fell silent and soon vanished. Through the remainder of the year, I saw no more of them. Probably they had descended to lower and warmer regions.

In contrast to certain other species of *Turdus* in the high mountains of Central America, the Black Thrush is a brilliant, if somewhat erratic, songster. Early one frosty morning in February, my attention was drawn by a song that I had heard once before but had not yet traced to its source. It consisted of the finest of thrush-like notes alternating with trivial and harsh sounds, a violent contrast of melody and dissonance such as is frequent in the songs of the Mimidae; I felt sure that the White-breasted Blue Mockingbird was their author. The notes floated down through the chill morning air from a ridge above the road, and led me up a steep slope, slippery with fallen oak leaves, through open woods of pine and oak. After a little stalking, I found the songster perching high in the top of a pine tree, and was more than a little surprised when my field glasses brought home to me the figure of a Black Thrush, for I was not then aware that any of the thrushes indulged in mimicry. His voice had a range equal to that of the Common Mockingbird. The most impressive notes in his medley, and the most conspicuous by their frequent repetition, were evidently his own birthright—notes so smooth and mellow, so full of the power to soothe and exalt the spirit, that to describe them was beyond my ability. But this minstrel of sublime genius did not scruple to descend to the meaner art of the mimic, and mixed his own incomparable phrases with many borrowed from less gifted musicians. His rendition of various harsh and mewing notes of Steller's Jay was perfect. He imitated the warbling of the Common Bluebird and the call of the Whip-poor-will and, I believe, attempted to reproduce the song of the Brown-backed Solitaire, but if so, without success. Interspersed here and there were many single notes—*peep*'s, *chuck*'s and the like—apparently the calls and flight notes of his neighbors.

As was to be expected of birds so versatile in their singing, not confined to inherited or traditional phrases as most songsters are, there was great individual variation in the Black Thrushes' performances. While most, if not all, of them introduced a certain number of harsh or merely chattering notes into their repertoire, some individuals used few of these in proportion to their own rich, melodious whistles, with the result that their compositions were of the highest quality and entitled them to a place among the best songbirds; but other individuals were more given to the utterance of unmelodious chatter, imitating chiefly the call notes of other birds, and continued their garbled hodgepodge, as inferior as that of the Catbird at his worst, until I tired of hearing it.

Some Black Thrushes had a song which consisted in running rapidly up the scale in a series of loud whistles. Although this exercise provided an opportunity to display their great vocal power, to my ear there was a disagreeable suggestion of flippancy in its tone. Of the notes borrowed from other birds, those of the Common Bluebird were the most frequent, and were generally rendered with such faithfulness to the original that, if it had not been for the medley of which they formed a part, I never would have suspected the true source of the notes. Next to these, various calls of Steller's Jay were perhaps most often copied. Even the peculiar *wake-up* and the rattling flight call of the Gray Silky-Flycatcher were not too difficult for this versatile mime; but he did not seem able to reproduce the Brown-backed Solitaire's unique chiming, nor the Yellow-backed Oriole's precise and mellow whistles.

BLACK-FACED SOLITAIRE

Myadestes melanops

The Black-faced Solitaire is nearly everywhere slate-gray, somewhat darker on its upper parts than below. A conspicuous black mask covers the forehead, sides of the head, chin, and upper throat. The wings and tail are also largely black. This elegantly plain attire is set off by the bright orange of the bill, legs, and toes. The eyes are dark. The female is similar to the male but may have slightly darker plumage, tinged with olive on the back. Both sexes are nearly seven inches in length.

This solitaire is found only in Costa Rica and western Panama, but it is closely allied to, and by some considered to be conspecific with, the Andean Solitaire of South America. According to Carriker (1910: 738), in Costa Rica it is "generally distributed over the whole of the country above 2,500 feet, even going down as low as 1,200 or 1,500 feet on the Caribbean slope." Slud (1964:298) states that the Black-faced Solitaire "inhabits the mountainous interior from about 2,500 feet in the subtropical belt to 7,000 feet in the lower montane belt. It is most abundant, in general, between 3,000 and 5,000 feet." In my experience, the solitaire lives chiefly at heights greater than these statements imply. On the Pacific slopes of the Cordillera de Talamanca, above the valley of El General, I have never found it as low as 3,000 feet in the breeding season; but from November to early February, when nights are coolest, an occasional bird will wander down through the woods as low as 2,500 feet, sometimes calling attention to its unexpected presence by singing a few notes. In the days when the oak forests on the upper slopes of the Cordillera were still scarcely touched by the axe, one could, in March, climb from 6,000 to 9,000 feet to the almost constant accompaniment of the solitaire's enchanting song. On the Cordillera Central I have found this thrush abundant between 5,000 and 7,500 feet, but it ranges beyond these limits in both directions.

In Costa Rica the *Jilguero,* as the solitaire is called, is famous as a songster and, in consequence, all too familiar as a caged bird; yet of its life in the free state little is known. This thrush takes full advantage of the excellent opportunities for concealment provided by the wet, moss-draped forests in which it dwells, so that it is difficult to watch while it forages and even while it sings. It persists in small patches of woods that have been allowed to remain, usually in deep ravines, amid the pastures and cultivated fields that relentlessly encroach on the heavy mountain forest that is its true home; yet here, too, the bird is difficult to observe amid densely tangled shrubbery and close-set stands of tall, cane-like bamboos. Occasionally a solitaire ventures forth from the forest to forage in adjacent pastures, but it does so far less than some other woodland birds, such as the Ruddy-capped Nightingale-Thrush, the Collared Redstart, and the Prong-billed Barbet. Black-faced Solitaires seem never to flock with others of their kind nor to accompany mixed parties of small forest birds. I have no evidence that they remain paired after the breeding season. True to their name, they are solitary creatures.

The Black-faced Solitaire eats a variety of berries, including those of the small palms that grow beneath the highland forests, and doubtless also insects. Although bird fanciers have carefully studied the food preferences of their captive *Jilgueros,* there are few observations on its feeding habits in the wild. Both Slud (1964:298) and I have seen a solitaire with army ants, in the lower parts of the species' altitudinal range. Shier than most of the birds that had gathered to snatch up the insects and other small creatures that fled from the raiding ants, the solitaire that I watched lurked in the background and soon vanished into the forest, whence, shortly afterward, I heard its song.

Voice

Individual Black-faced Solitaires vary greatly in their musical ability. It is evidently difficult to achieve perfection in the delivery of a song so complex and exquisitely modulated as theirs, and the inferior songsters are probably in most instances young birds whose performances will improve with practice. On a morning toward the end of March, 1938, I had the good fortune to watch one of the more accomplished musicians sing while perching on a naked horizontal length of a long liana amid the forest. Fully to appreciate the melodies of so careful and finished an artist, it was necessary to listen at close range and with undivided attention. I stood long in his presence while he sang; and in this wild mountain forest where man was still scarcely known as a formidable enemy, he seemed undisturbed by the proximity of his solitary audience. The carefully turned phrases, the intricate modulations of his voice, the delicate shades of tone, formed a composition of exquisite beauty which held his auditor spellbound. While listening to this solitare's song, I found it hard to resist the impression that he was a selfconscious artist striving earnestly to attain perfection in his art, and carefully noting the effect that he produced. Sometimes he seemed to go astray and sound a false note. Yet for all the superb beauty of his music, his voice was not notably rich or full; many an inferior songster has a sweeter and mellower voice. He was a musician whose genius enabled him to produce divine melody with an instrument somewhat deficient in tone.

To hear the Black-faced Solitaire's song with the songster in full view is an unusual experience, for mostly he performs well hidden amid the trees. The sweet, unhurried, mellow flute notes, floating down from a solitaire invisible amid the foliage, seem the utterance of the forest's disembodied spirit, expressing its multiform beauty, its calmness, and that sweet melancholy which clings to all things mutable. One

FIG. 4. Forest of oaks and other broadleafed trees at about 7,000 feet at the western end of the Barba massif in Costa Rica's Cordillera Central, habitat of the Black-faced Solitaire, Mountain Thrush, Yellowish Flycatcher, Barred Becard, Black-cheeked Warbler, Violet Sabrewing, and other birds of the subtropical zone. 21 June 1963.

March, years ago, I climbed from the valley of the Río Buena Vista to the summit of El Cerro de la Muerte with the pensive whistles of the solitaire so continuously in my ears, for three thousand vertical feet, that it was easy to imagine that a single songster followed unseen to cheer my toilsome ascent. Undoubtedly I passed a succession of solitaires; but I saw none until late in the afternoon, when I noticed two slate-gray birds, with black faces and bright orange bills and feet, perching close together on a branch above the steep, rocky trail. Not at all shy, they permitted me to see them well before they flew off, still keeping together.

The Black-faced Solitaire's song defies analysis, at least by one who is no expert in musical notation. Once, however, I tried to describe one of the simpler versions. It consisted of a liquid, undulatory phrase followed, after a brief pause, by two slight notes and a prolonged mellow whistle. This solitaire sings with a calm deliberation that seems the product of a studied art and contrasts strongly with the wild, spontaneous exuberance of the singing *Guarda Barranco* or Brown-backed Solitaire of northern Central America. In his style of singing, the Black-faced Solitaire more closely resembles the Slate-colored Solitaire of northern Middle America; yet the songs of these two species are also quite distinct.

The Black-faced Solitaires sing most in March, when in the wilder forests above 5,000 feet in the Costa Rican mountains, through most of the day, one is rarely beyond hearing of their voices. They sing through much of April, but in May, when parents are busy with young, they rapidly fall silent. From June to August or September they are rarely heard; but in October, when they have probably finished molting, they sing freely again. At Montaña Azul in the wet and gloomy month of October, 1937, I considered them the most songful of the birds. I doubt that this autumnal renascence of song was associated with nesting. There was a distinct decline in the amount of singing during the following months, but I heard solitaires occasionally until they came into full song in March.

Nesting

It is strange that so little has been written about the nest of a bird so well known to the Costa Rican mountaineers as the *Jilguero*. The only published description that I have seen is by Blake (1956). In 1938 I found a solitaire's nest, and 25 years later two more were shown to me. The first nest, which I discovered at an altitude of 5,600 feet at Montaña Azul, was 10 feet above the ground, in a furrow in the side of an irregular leaning trunk at the edge of low second-growth woods, beside a pasture. The gnarled trunk was thickly overgrown with mosses, ferns, and other epiphytes, including a large cluster of orchids that grew on a small branch above the nest, shielding it above and keeping out the rain. Situated in a dark nook and composed largely of green materials that blended with its setting, this nest was so well hidden that I should not have noticed it if a solitaire had not flown out as I passed by. After the nestlings took wing, I removed the structure for examination. It was a shallow cup, made chiefly of a great mass of green foliaceous liverworts. It was thickly lined with black fibrous roots, with an admixture of the dark, slender stems of mosses and liverworts from which the leaves had decayed, although a few of these plants with filamentous stems still bore minute leaves.

The other two nests that I have seen were in niches in nearly vertical mossy banks, beside cowpaths that led up the sides of deep wooded ravines that intersected extensive pastures at La Giralda, about 7,000 feet above sea level. The banks were respectively 2 and 4 feet high, and in each case the niche containing the nest was near the top. Both of these nests were bulky open cups composed almost wholly of green moss, with a lining of blackish rootlets and similar materials. The interior of one of these nests was about 3 inches in diameter by 2 inches in depth. Both nests were roofed with fallen leaves, chiefly from the tall bamboos that grew densely in the ravines. I could not tell whether the leaves had simply fallen here or the solitaires had made an effort to provide a roof for their nests.

The first nest contained three newly laid eggs on 30 April 1938. These eggs were white, heavily mottled all over with bright rufous-brown, with the pigmentation densest on the broader end. They measured 25.8 by 19.1, 25.8 by 19.1, and 25.0 by 19.1 mm. When I was shown the second nest on 1 May 1963, a fledgling shot out and flew strongly across the ravine. I was told that there had been two eggs and nestlings, one of which had flown a day or two earlier. The third nest contained three newly hatched nestlings on 26 May 1963.

To my regret, all of these nests were so situated that it was hardly possible to set a blind where it commanded a satisfactory view of them. At the first nest, all three eggs hatched 12 or 13 days after the last was laid. The newly hatched nestlings had pink skin with sparse, dark gray down. The interior of the mouth was yellow, and the flanges at the corners were whitish. From the first nest, the three nestlings departed when 16 days old; at the third nest, the nestling period for the three young that were successfully reared was 15 or 16 days. With their nearly black bodies heavily spotted with deep buff on the upper parts and breast, the fledglings were strikingly handsome. The many regurgitated seeds on the bank below their nest indicated that berries had entered largely into their diet. I never saw a solitaire simulate injury in a convincing fashion; but as I approached one of the nests on a bank a day or two after the nestlings hatched, a parent emerged from it and flew slowly down the cowpath ahead of me with widely spread, fluttering wings, each of which displayed a narrow white bar not usually visible. She alighted on a low branch projecting over the path, and when I drew nearer she flew down into the bamboos that formed a dense thicket in the ravine.

BROWN-BACKED SOLITAIRE

Myadestes obscurus

The Brown-backed Solitaire is a stout, dull-colored thrush about eight inches long. In both sexes, the whole top of the head, the hindneck and sides of the neck are slaty gray. The back and rump are olive-brown. The central tail feathers are brownish gray; the others are blackish, with dull white tips on the two outermost on each side. There is a conspicuous white ring about each eye. The cheeks are dusky, with whitish flecks. The dull white of the chin and upper throat merges into gray on the breast, which in turn pales to dull white on the abdomen. The bill is black; the eyes brown; the legs and feet dark.

The Brown-backed Solitaire ranges through the mountains from

Mexico to El Salvador and Honduras. In northwestern Mexico, it winters in the foothills and lower mountains and breeds up to 8,000 feet. In more southerly parts of Mexico, it appears to extend only slightly higher (Miller *et al.*, 1957:195-196). In Guatemala, where the solitaire resides from about 5,000 to 10,000 feet above sea level, it is known as the *Guarda Barranco*. Almost every Guatemalan is familiar with it by name and reputation; but many more, I suspect, have seen it in cages than in its woodland home. The name, which means "Guardian of the Ravine," sheds more light on the extent of deforestation in the country than on the habits of the bird. I doubt whether the solitaire has a special predilection for ravines; it is simply a bird of the forest; but since, in all the more thickly settled parts of the country, forests dense enough for it are restricted to ravines too deep for cultivation, these are the places where it is usually found. On the Sierra de Tecpam, fortunately, the forests were by no means confined to gorges; and so I knew this solitaire merely as a bird of the deep woodland, and never came to associate it closely with profound, inaccessible *barrancos*.

The specific name of this *Myadestes* was happily chosen. Not only is it obscure in plumage; it likewise lives obscurely, rarely leaving the shelter of the heavier forests, where there is sufficient undergrowth for its concealment. Dry, open, second-growth woods are not to its taste. Even when the male sings, he is careful not to be seen; he perches amid the foliage, instead of on the treetops like the Rufous-collared Thrush, and he does not become too absorbed in his musical outpouring to remain alert and difficult to stalk.

I have included the Brown-backed Solitaire in this book chiefly for its song. But how shall I set about to describe a song so utterly unlike all familiar bird songs, and how can I convey the reverential feelings that it inspires as one stands listening to it amid the dark mountain forests? Doubtless it bears more resemblance to the song of one of the more typical thrushes than to the song of a finch or a wood warbler; yet it is so different in character from the other thrush songs known to me that, when I first heard it, I was by no means certain that its author was a member of this family. The *Guarda Barranco*'s music is a chiming of sweet-toned pipes, a wild, jingling melody that, heard from a little distance, fills the listener with a profound sense of the freedom of the forest and its mystery. I say "heard from a little distance" because that is the way this song should be heard; to fall completely under its spell, one must stand not too near the singer. Why should not certain bird songs have been developed to be heard from afar; just as a carillon, which is said to make a most intolerable din in the belfry, is designed to produce its most pleasing effect at a distance? To a listener immediately below a singing Brown-backed Solitaire, there is a suggestion of sibilance, of hollowness, in the notes, as though the bird were blowing through a pipe that he had not enough wind to fill. The first notes of the song have a very ventriloquial effect, as though the songster were far away; as the music continues, the hearer is surprised to find him amid the foliage nearly overhead.

On the Sierra de Tecpam in 1933, these solitaires sang most in February and March; in April their singing had definitely begun to wane, and after the return of the rains in May they fell into a silence which they rarely broke through June and July. In August they recovered their voices and sounded their wild music now and then through the remainder of the year—the only thrushes which sang on this mountain during this period. In the wet season of the following year, which was far less stormy, I heard the solitaire sing more frequently as I traveled on horseback through the mountains of western Guatemala. According to Wagner (1955), in Mexico Brown-backed Solitaires sing throughout the year, even occasionally while they are molting in August and September. But even when not in song, the Brown-backed Solitaire is a consummate musician, if I may be pardoned the paradox. Its call note is beautifully clear and round. On a misty evening in June, I listened while a pair of these birds, unseen in the woods, called and answered each other over and over. The clear whistles, coming now from this side, now from that, made the finest music, although it was composed of only a single reiterated note.

I did not have the good fortune to find the solitaire's nest. Early in March, an Indian showed me the mossy beginning of a nest, which he affirmed belonged to the *Guarda Barranco,* well hidden amid herbaceous vegetation in a little depression in the ground at the top of a roadside bank, in the woods. But this nest was never completed, the birds having doubtless discovered that they had selected a location too public for their liking; and in spite of considerable searching, I never found another.

In Mexico, however, Wagner (1955) found 27 nests between March and July. In the cloud forests of the Sierra Madre of Chiapas, breeding begins at the height of the dry season in March, as evidently it also does in the western highlands of Guatemala; but in the mountains near Mexico City, the solitaires begin to build after the first showers of May or June. The nest, an open cup of moss lined with straws or pine needles, is placed in a variety of situations: in niches in earthen banks; in unfinished burrows dug by motmots; in crannies in rocky cliffs; beneath and among the exposed roots of old trees; and even on the surface of the ground. The set consists of two or three eggs, which are glossy brownish white uniformly flecked over the whole surface with sepia. The incubation period was not determined, but the nestling period was about 17 days. Because of the solitaire's habit of catching insects in the air like flycatchers, and its nuptial song flight, Wagner questions the validity of the present classification which includes *Myadestes* in the thrush family. But the solitaire that I know best, the Black-faced, seems to me a sufficiently thrush-like bird.

Guatemala has a second species of solitaire, the Slate-colored; and although I have been only a transient in its range, I cannot pass it by unnoticed. The two solitaires divide the country between them in an interesting fashion. The brown-backed species lives all over the central highlands, up to at least 10,000 feet, and on the Pacific slopes of the great volcanoes down to 5,000 feet or possibly a little lower, where I sometimes heard it at the upper edges of the huge coffee plantations on this slope. The Slate-colored Solitaire dwells in the heavy mountain forests of the Verapaz and northern El Quiché, chiefly at altitudes below that to which the Brown-backed Solitaire descends.

Riding down the steep northern slope of the Cerro Putul in the department of El Quiché, one cloudy afternoon late in November, I heard for the first time the sweet music of this incomparable musician issuing from his native forest. Enchanted with the effect, I at once expressed to my companion the hastily formed conviction that the *Jilguero,* as this bird is called, sang better than the *Guarda Barranco;* but he, a native Guatemalan, did not agree with me. As we continued our journey, much of the time through unbroken forest, I had ample opportunity to become familiar with the Slate-colored Solitaire's song, to "let the novelty wear off"; and after a while I was no longer sure that I agreed with myself. In fact, the songs of these two solitaires differ so radically that to compare them is not quite fair. I loved the music of the Guardian of the Ravine for its wild, unstudied exuberance, which expresses the untamable spirit of nature, lingering unsubdued in wild, inaccessible ravines after all the surrounding lands have been shorn of their woodland splendor and planted with coffee or corn. I admired the song of the *Jilguero* for the studied perfection of its exquisitely modulated notes, the melting sweetness of its tone. This solitaire is a careful and formal artist, who seems, despite the remoteness of his mountain home, to have found the opportunity to take lessons in a conservatory. His cousin of the highlands is a wild, untutored mountaineer, careless of studied art, but overflowing with joyous vitality that must find vocal expression.

Family CYCLARHIDAE

RUFOUS-BROWED PEPPER-SHRIKE

Cyclarhis gujanensis

The Rufous-browed Pepper-Shrike is a stout, short-tailed, strong-billed bird about five and a half inches long. The top of its head is brownish gray, bordered by broad superciliary bands of cinnamon-rufous which sometimes meet on the forehead. The sides of the head and hindneck are light gray. The rest of the upper parts are olive-green. The under parts are largely yellow, usually tinged with olive on the throat and chest, and fading to white on the abdomen and under tail coverts. The notched upper mandible of the short, thick, laterally compressed bill is pale brown or flesh-colored, and the lower mandible is grayish blue with a pale tip. The eyes are orange-yellow; the legs and feet flesh-color. The female closely resembles the male but is sometimes slightly duller. This description refers to the Costa Rican race, *C. gujanensis subflavescens.* Other races of this widespread species differ slightly, especially in the color of the iris, which ranges all the way from yellow through orange-yellow to reddish brown and tawny-ochraceous.

The Rufous-browed Pepper-Shrike is found from northeastern Mexico to Peru, Brazil, and northern Argentina. Throughout most of this vast territory, it inhabits warm lowland forests, including those of the Pacific coast of Central America, from the Gulf of Nicoya northward. But in certain regions it is confined to the highlands, as in central and southern Costa Rica, where it lives chiefly between 3,000 and 6,000 feet above sea level. On our farm at 2,500 feet on the Pacific slope of southern Costa Rica, the pepper-shrike is so rare a visitor that years may pass without my becoming aware of its presence. A thousand feet higher, its ringing song may be heard through much of the year. On the Sierra de Tecpam in western Guatemala, at an altitude of about 8,500 feet, I first met the species in early October, more than nine months after my arrival in these mountains. In Guatemala the pepper-shrike occurs chiefly from 1,200 to 6,000 feet (Griscom, 1932:322), and the few that I saw at far greater heights were probably wandering after the close of the breeding season.

In the tall forests at middle altitudes in Costa Rica, where chiefly I have watched the pepper-shrike, I have most often noticed it high in the great forest trees. At least, here is where it nearly always sings and calls attention to itself; although to forage and to nest it often, perhaps habitually, descends to lower levels of the vegetation. But even in Costa Rica, the pepper-shrike is by no means confined to the lofty subtropical forests. It is not rare on the Central Plateau, where it sings in the taller shade trees about dwellings, in the coffee plantations, and along the roadsides, and on occasion may be heard proclaiming itself loudly even in low thickets. Likewise in Mexico, Venezuela, and other regions, it is often found in the shade trees around houses. In the many parts of its range where it inhabits semi-arid country, the pepper-shrike must perforce adapt itself to low and rather open woodlands. In still drier regions, where cacti and thorny scrub prevail, the pepper-shrikes prefer the green trees and shrubbery along the watercourses (Todd and Carriker, 1922:430). In Trinidad, according to Belcher and Smooker (1937:516), pairs keep together throughout the year. The little evidence that I have gathered on this point in Costa Rica inclines me to believe that here the pepper-shrikes are solitary during the second half of the year, but further observations are needed.

Food

Although they prefer the high treetops for singing, pepper-shrikes sometimes drop down to lower vegetation to forage. One day, after I had listened gratefully to some pepper-shrikes singing profusely in the roof of the forest far above, I saw two of them fly down to hunt amid the tangles of vines that covered some low, second-growth trees across the trail from where we sat eating our lunch. They continued to sing as they foraged amid the dense foliage, which most of the time screened them from view.

The first pepper-shrike that I ever saw was hunting food among the lower boughs of trees in the high mountains of Guatemala, with a mixed flock of small birds consisting chiefly of Townsend's and Crescent-chested Warblers. It moved slowly and deliberately, substituting careful ocular inspection of the surrounding foliage for the incessant agitation of its warbler companions. Whenever the pepper-shrike found a fairly large caterpillar, it held the larva beneath a foot while it tore off pieces with its bill and swallowed them—a mode of eating which it shared with the Chestnut-sided Shrike-Vireo of the same region. True vireos follow this method of dismembering their food less frequently, and many small passerines seem never to use a foot for holding things. On the few subsequent occasions when I have succeeded in watching pepper-shrikes devour their food, this consisted chiefly of fat caterpillars, which they tore apart just as my first pepper-shrike had done. Once I saw a pepper-shrike extract a pupa, evidently of a moth, from a cocoon, tearing the tough fabric with its powerful bill while it held the silken case beneath a foot. If such items are frequently included in this bird's fare, the usefulness of its remarkably heavy bill would be apparent.

Voice

Nearly everyone who has written, even briefly, about the Rufous-browed Pepper-Shrike has had words of praise for its song. On the Isthmus of Tehuantepec, long ago, I recorded that it had a sweet song, more like a wood warbler's than a vireo's; but the songs of more southerly races are certainly not warbler-like. In Trinidad, according to Belcher and Smooker (*loc. cit.*) "there is no season at which one may not hear, from a high leafy tree, the loud call which seems almost to ask for 'translation,' and has been rendered variously 'Do you wash every week' (stresses on 'wash' and 'week') and 'We're waiting to hear you' (with 'wait' and 'hear' accented)." The pepper-shrike's song combines in a remarkable fashion volume and carrying-power with clearness and sweetness of tone. With vireo-like persistence, he sings the same song over and over until he tires of it—and sometimes his audience does, too, despite its beauty—then takes up another refrain and treats it in the same manner.

As Belcher and Smooker so justly remarked, these songs almost ask for "translation"; few birds' songs that I know are so consistently easy to paraphrase. *We don't believe it* and *We've been wishing to meet you* are renderings which spontaneously suggested themselves to me; while *Río Chirripó* was a persistent refrain of pepper-shrikes who lived within hearing of that sonorous mountain torrent of southern Costa Rica. In this country, most of the songs consist of five syllables, rarely more. It is not only into English that the pepper-shrike's verses are easily translatable; according to Mitchell (1957:194), Portugese-speaking Brazilians translate one of its songs as *Gente-de-fora-já-chegou*, which, translated once more, means "Strangers have just arrived." By this onomatopoeic name the bird is known to the people.

In Costa Rica, I have heard the pepper-shrike's mellow, far-carrying song in every month of the year; although in a single district there may be a period of several months, centering around the autumn equinox, when it is rarely heard, probably because the bird is molting. It sings freely not only at the beginning of the dry season in December and January, its solitary voice breaking the perplexing, almost ominous, silence of the lofty mountain forests on days when sunshine and flowers seem to invite every bird to sing; but its mellow verses, ringing out from mist-drenched woodlands, help alleviate the gloom of the wetter months. The pepper-shrikes' vocal performances are most impressive when two rivals sing against each other, their clear notes sounding now on this side, now on that. Apparently these birds, which never flock with others of their kind, defend territories throughout the year. Since pepper-shrikes are even harder to detect amid the foliage than some kinds of vireos, their presence in many localities would be overlooked if they did not sing so loudly and persistently.

In early August of 1964, I went to live near Escazú, on the Central Plateau, while I taught ornithology at the University of Costa Rica. From time to time a puzzling bird call issued from among the trees surrounding our cottage. It consisted of from four to eight similar notes, the first rather high and shrill, each following note slightly lower, the whole falling sequence delivered in a most peculiar tone. I suspected that the author of this strange utterance was a vireo; but since it was heard at very long intervals—perhaps only two or three times in a day—and between calls the bird remained hidden in silence amid the foliage, I could not for a long while identify it. But one morning in September, after I had been wondering about the identity of the caller for over six weeks, two of these baffling birds were nearby. There were answering calls from the roadside trees, and one bird chased the other. Then, stimulated by this encounter, the nearer bird called a number of times, until finally I succeeded in watching it in the act. It was, to my great surprise, a Rufous-browed Pepper-Shrike. In October and November, it continued to call in the vicinity, but more seldom.

Strangely enough, during four months' residence in this locality, we never heard the familiar song of the pepper-shrike. Yet in August and September a pepper-shrike sang much on the campus of the University a dozen miles away, and here we never heard the peculiar call of falling notes. I suspect that this is the utterance of the female. It seems likely that the sexes separate after the breeding season; male and female hold separate feeding territories; the male proclaims his ownership with his loud ringing song; at long intervals the female asserts her claim by uttering her less musical phrase just once. I lack sufficient evidence to prove this theory, but it is the one which best fits my observations on these elusive birds that are heard so much more often than they are seen.

Searching through old journals, I found that over 30 years earlier, in the Guatemalan highlands, the first pepper-shrike that I ever saw called "only once, when it uttered in rapid succession about seven loud whistles, with just enough of a screech to impart a distinctive flavor to the refrain." This evidently was the call that I attribute to the female. Probably this is also the "quite different call which reminds C. F. B. of one of the calls of *Tschagra senegalus* in East Africa," mentioned by Belcher and Smooker (1937:516).

Nesting

Great was my elation when, on 27 March 1938, I discovered my long-sought first nest of the pepper-shrike, in the Costa Rican mountains near Vara Blanca at an altitude of 5,600 feet. In form and situation, it was a structure such as a large vireo might have made. The deep, hemispherical cup, open above, was fastened by its rim to the arms of a fork near the end of a branch, 16 feet up in the top of a small, vine draped tree that grew in a bushy clearing near the forest's edge. My mirror, raised up on a long stick, reflected one egg in the bottom of the suspended pouch. A second egg was present on the following day. As I could not reach the nest without disturbing its surroundings, I never touched the eggs, but contented myself with observing in the mirror that they were white, sparsely speckled on the thick end with small dots of bright brown. Five days later, to my dismay, the nest was empty, pillaged by some undetermined predator.

I consoled myself for my loss by cutting down the nest for closer inspection. It measured 2½ inches in internal diameter by 1¾ inches in depth. The fabric was very thin and open, permitting much light to pass through its meshes. The exterior was composed of long, slender, profusely branched lichens of a light gray color and a few pieces of green moss. These materials were bound together, and to the supporting twigs, by liberal wefts of tawny silk from spiders' egg cases. The spiders' eggs had been carried to the nest along with the silken cases; some of them had hatched, liberating little black spiders which crawled over the nest. The interior was lined with rather coarse, wiry vegetable materials, including the slender rachises of fern leaves, coarse bast fibers, and fibrous roots, doubtless of epiphytes.

At the end of May, I found a pair of pepper-shrikes building a nest in the same locality. They had placed it among the vines that draped the lower part of a tall dead trunk standing in a narrow clearing in the forest. Attached near the end of a branch of the vine, it was about 30 feet above the ground. Although inaccessible to me, from the ground I could see that it resembled the first nest in form, and that it was made of the same gray, branching beard-lichens (probably a species of *Usnea*), bound together with a liberal amount of cobweb. When I first noticed the nest, the builders were already gathering coarse fibers for the lining. Both sexes brought materials, and each placed its own contribution in the nest and helped to shape it. The two always flew up together, alighted in the lower part of the vine tangle, then hopped and flitted slowly upward to the nest. One waited nearby while the other arranged what it had brought. On nearly every visit, the male sat in the nest to arrange its materials, even if he had brought nothing himself; probably he had come with empty bill because his partner had found suitable material more quickly than he and he felt constrained to accompany her to the nest even if he had nothing for it. When the male took his turn in the nest before his mate, he waited, singing, close by while she worked at it. After both had deposited and arranged their contributions, they flew off together for more.

This nest was finished a few days later. At 6:20 a. m. on 6 June, I watched the male enter it to sit for an hour, during which he sang often and loudly. At the end of this interval, he left as his mate approached. She sat for only 4 minutes, then jumped out and hopped through the surrounding vines for an equal period, and then returned and sat for only three minutes more. I could not learn whether eggs had been laid, nor why the male sat so much longer than his partner.

Two days later, on 8 June, I began at 5:50 a. m. to watch this pepper-shrikes' nest. It remained unattended until 6:08, when the male came to sit in it. Here he stayed for nearly 2 hours, repeating his song at intervals of a few seconds in his loudest, clearest voice. He did not cease singing until his mate arrived at 8:02, when he made way for her. She flew off without sitting. Then neither partner came near until 9:22, when the male returned and sat for 10 minutes, singing as loudly and constantly as before. After his departure, I waited for 20 minutes more, but neither of the birds returned to take charge of the nest. On numerous later visits, I always found it unattended. Had the female failed to lay, or had the male stubbornly persisted in sitting in a nest from which the eggs had been stolen, vainly trying with his song to persuade his mate to join him in this useless occupation? Because of the inaccessibility of the nest, I could not answer these questions. Years later, I watched a female Gray-headed Tanager "incubate" in an empty nest as assiduously as though she had eggs.

About the same time as these pepper-shrikes built among the vines, I found another pair finishing a nest, 30 feet up in a shrub of the epiphytic heath *Cavendishia* growing on a trunk that stood in a pasture with scattered trees. After completion, this structure, which likewise I was unable to reach, appeared to be abandoned. Could it be that the breeding season was then waning, and the females were unable to produce eggs for these late nests?

In the Orinoco region of Venezuela, Cherrie (1916:158) found a nest of the pepper-shrike on 23 May 1907. "The nest was situated in a *Chaparo* oak that stood near the edge of an extensive open savanna. It was placed at the extreme tip of a long horizontal limb, about 4.5 m. from the ground, suspended between forked twigs. For a pendant nest it was unusually shallow; the walls thin, and it might be described almost as a net woven between the forks and sagging in the center. Outwardly it was composed entirely of soft grasses, and there was an inner lining of a very few hair-like vegetable fibers . . . The nest walls were so thin and the meshes so open, that the eggs were visible when looking from the ground through the bottom of the nest." This nest contained three fresh eggs, which were "white, faintly washed with buffy pink and marked with blotches, spots and tiny dots, varying in color from hazel brown to dark chestnut." The male was incubating them.

Evidently the Rufous-browed Pepper-Shrike builds its nest of quite different materials in different parts of its wide range. Two nests found in Trinidad and attributed to this species "were deep cups of black horsehair-like fiber with turned-in rims by which they were slung. They were so thinly built that they could be seen through from below." These nests were suspended in forked twigs of bamboos about 30 feet up (Belcher and Smooker, 1937:516). A nest found by Worth (1938) at 4,100 feet elevation in the Province of Chiriquí, western Panama, was only 7 feet above the ground in a coffee bush in a plantation. Constructed largely of a type of moss that grew luxuriantly on the neighboring forest trees, it was more heavily and clumsily woven than the nests of true vireos.

This nest found by Worth held, on 3 July 1937, two nestlings with opening eyes and sprouting pinfeathers. They bore no trace of natal down and had evidently been quite naked when they hatched. A brooding parent remained on the nest until almost touched by a hand, but otherwise the parents showed little concern when their nestlings were visited. They were seen to bring only insect food, principally soft caterpillars. When their plumage expanded, the young resembled the adults.

Family VIREONIDAE

YELLOW-WINGED VIREO

Vireo carmioli

Scarcely four and a half inches long, this little vireo has greenish olive upper parts. The wing and tail plumes are dusky, with pale olive edgings, and there are two broad, pale yellow bands on the coverts of each wing. A pale yellow superciliary line extends from the lores to a little behind each eye, and the lower eyelid bears a spot of the same color. The sides of the head and neck are greenish olive. The underparts are light yellow, dullest on the throat, brightest beneath the tail. The short bill, the legs, and the feet are blackish; the eyes, brown. As in other vireos, the sexes cannot be distinguished by their appearance.

The Yellow-winged Vireo is restricted to the highlands of Costa Rica and neighboring parts of Panama, where it occurs from about 5,300 feet up to at least 9,000 feet. It lives in the heavy, epiphyte-laden highland forests, where it usually forages high in the great trees but occasionally descends to the lower boughs or the undergrowth, and it also inhabits neighboring pastures and other clearings with scattered trees. It often joins the mixed flocks of wood warblers and other small birds that wander through the mountain forests after the close of the breeding season. One or a pair of Yellow-winged Vireos may keep close company with one or two Flame-throated Warblers as they hunt through the foliage of alder trees in a pasture, the two kinds of birds flying together from tree to tree. I gathered no evidence that these vireos, which are permanent residents of the Costa Rican highlands, remain in pairs during the second half of the year.

When foraging, these vireos flit from twig to twig, on each of which they pause while they turn their head or their whole body from side to side to scan the surrounding foliage. They pluck insects and caterpillars from bark as well as leaves. Sometimes they make a short dart to snatch something from the foliage while hovering on wing. Their deliberation contrasts with the more rapid, unresting movements of the warbler who often accompanies them. As far as I have seen, these vireos eat only insects, spiders, and the like. (In including this vireo among the thicket foragers, Hamilton (1962) based his judgment on inadequate information.)

Voice

The song of the Yellow-winged Vireo is typical of its family, consisting of short phrases separated by longer pauses, but it is delivered with a most peculiar intonation, which makes it easy to distinguish from the songs of other species. The tone is often a high falsetto, not unpleasant to hear, which gives the impression that this little vireo is parodying his larger cousins. He sings his mocking verses less continuously than the Yellow-green Vireo or the Red-eyed Vireo, usually high in a tree where the performer is difficult to find. In May one sang *vire—chichui—chuyee,* and also *viree—viree—cheeyu.* Another seemed to say *viree—witchum—viree—witchum. . . .* Eisenmann (1962) noticed that the Yellow-winged Vireo's voice had some of the huskiness characteristic of the much larger Yellow-throated Vireo. At La Giralda in 1963, I heard this vireo sing from eary March into July. Twenty-five years earlier at Montaña Azul, it sang from August into November. A little sunshine would encourage it to raise its voice even after a week of almost continuously wet weather on this storm-beaten mountainside. Evidently this vireo sings through most, if not all, of the year. At two nests, I heard only the male sing; the female seems to be songless. While building, a pair uttered very low, soft notes.

Nest Building

The only two nests of the Yellow-winged Vireo of which I know were situated in trees growing amid pastures, at an altitude of about 6,700 feet at the western end of the massif of Volcán Barba. The first, found when newly begun on 21 May 1963, was 9 feet up in a small Winter's bark tree (*Drimys Winteri*), one of a row of these trees. The second, 25 feet up in a middle-sized alder tree (*Alnus acuminata*) that stool alone, held nestlings on 24 June of the same year. This nest was inside the tree's crown, well screened by foliage; the first was near the end of a thin projecting branch in a far more exposed situation. Each nest was hung between two slender, horizontal, diverging twigs, in typical vireo fashion.

When I found the first nest at about 6:40 a.m. on 21 May, the vireos were just starting to build. They had chosen for the attachment of their nest two slender, diverging branchlets which sprang from a thin upright branch, one of them an inch above the other. But the upper branchlet drooped more than the lower one, so that a short distance from the vertical stem the supporting twigs were on the same level. The vireos were plucking cocoon silk, cobweb, bits of lichen, and other fragments from the bark of neighboring trees and taking them to this exposed site. They tried to wrap the silk around the smooth branchlets, but some of it slipped off and was lost. To my great surprise, they also tore pieces from living green leaves and carried them to the nest site; I had never seen a vireo use such material in its nest. Both sexes were actively engaged in this undertaking; but since the male rarely sang while he worked and I had no other means of distinguishing the two, I could not tell which, if either, took the leading part. The two together made about 14 trips to the nest, then went away. After an interval, they returned and visited the nest twice. After another absence, they came back and took material to the nest 14 times. Their next spell of work included seven trips to the nest. Thus, in the hour from 6:40 to 7:40 a.m., they made a total of 37 trips to the nest. In the next 10 minutes, 7:40 to 7:50, they made a further 14 trips to the nest. Although the vireos worked without singing, I heard a few songs while they were off in the distance, foraging. While building, they uttered only very low, soft notes.

All this occurred while I had been standing in the roadway, 10 yards from the nest. Now I slipped through the barbed wire fence into the pasture where the nest tree stood and walked toward it. As I approached the nest, the male vireo also arrived, singing, with yellow cocoon silk in his bill. Despite my nearness, he went to the nest to place his contribution. Fearless of me, the pair continued to work while I stood with my head only 4 or 5 feet from them. While I remained here, they visited the nest six times in 10 minutes.

When I left the vireos at 8:10, they had attached a cable of cocoon silk between the two supporting twigs, to each of which it was anchored at a point about the vireo's own length from the upright branch on which these twigs were inserted. In this cable, which drooped loosely and swayed in the breeze, were entangled tufts of silk, small pieces of gray lichen, and scraps of green leaves, the largest about an inch long. Between this cable and the upright branch, in the space that the future nest would occupy, there was nothing at all. By midafternoon of the same day, the cable had been widened to a band about 2 inches across, which hung between the twigs in a catenary loop. To form this band, the original cable had been broadened inward rather than outward, and the inner edge of the band was now about 2 inches from the bases of the supporting twigs.

By 8:50 of the following morning, 22 May, the band joining the supporting twigs had been still further widened, so that it now resembled a little hammock, with its free edges higher than the middle. In the next hour, the vireos brought contributions to the nest at least 36 times—probably more, as doubtless I missed some of their visits while watching them gather their materials. When they were working most actively, one would usually reach the nest before its partner left, so that for minutes together a vireo was always sitting there, arranging the materials. The two seemed to be taking fairly equal shares in the task, but there were indications that the male did somewhat more than the female: on at least four occasions, one bird came alone with material, placed it in the nest, and sang a few times before flying beyond my view. While the two worked together, I heard no song and could not distinguish the sexes.

When the two partners worked together, they as a rule did not go far for their materials. I saw them tugging hard to detach pieces of lichen, both foliaceous and branched, from the bark of neighboring trees. They, or one of them, went repeatedly to a low shrub with thin leaves that grew near the nest and tore pieces from its foliage. When the vireo happened to pull off a fairly large fragment of leaf, the piece was promptly dropped. But the bird retained smaller bits in its bill while it tore additional shreds from the leaf, and it flew to the nest with several pieces of the green tissue. The foliage of this shrub was badly frayed by the building birds.

Much silk of different shades, chiefly whitish and yellow, was brought to the nest, mostly from sources beyond my view. The vireos spent much time perching on one of the supporting arms, seizing tufts of silk in their bill, and moving them across to the opposite arm, thereby stretching strands of silk from side to side of the growing nest. Rarely one of the builders would sit for a few seconds in the hammock, making shaping movements; but nearly all the arranging of materials was done from the sides. During most of the hour that I watched on this morning, a strong wind swayed the nest tree and the nest; but the vireos worked hard in spite of this disturbance.

By 23 May, the free sides of the nest (the inside toward the center of the tree and the outer side) had been built up almost to the height of the attached sides, converting the hammock into a little cup suspended by its rim. The outer surface seemed to be completed, but there was still no lining.

On 24 May, from 6:40 to 7:40 a.m., the two vireos brought material to their nest at least 48 times. All but one of these visits were made in the first 50 minutes of this hour, when the birds followed each other at intervals of about one minute. This energetic building went on despite the boisterous wind that whipped their nest and the surrounding foliage, making observation difficult, and the fine drizzle which this wind drove through the trees. The vireos brought the same kinds of materials as on the preceding days, including much cobweb or cocoon silk, lichens and moss plucked from bark, and fragments of green leaf. Much of this material was simply dropped into the cup while the bringer stood on the rim; but from time to time a vireo would sit in the cup to make vigorous shaping movements that shook the whole fabric, sometimes continuing this for a minute or more. But more often than they entered the nest the builders stood beside it, as on past days, and spent much time spreading cobweb over the supporting twigs and even over the outside of the structure. Frequently they stretched the cobweb from one side of the nest to the other, as at earlier stages of construction. I heard only one song during the hour.

By evening of this same day, 24 May, the size of the nest's cavity had been greatly reduced by the materials which the vireos had been dropping into it. The free sides of the rim were now fully as high as the attached sides, and the structure seemed to be finished or nearly so. Nevertheless, the vireos were still bringing a little more material at noon of the following day, 25 May. The walls had now become quite thick. Although no egg was laid until 29 May, four days later, by far the greater part of the construction of this elaborate nest had been done in five days of hard work.

This nest was in a position difficult to reach, and I decided not to examine it closely until after the young had flown. Unhappily, before this could occur, some predator, or perhaps some mischievous boy, took the eggs and badly tore the nest. But even in the mutilated state in which I finally held it in my hands, it impressed me as one of the most marvelous avian fabrics that I had ever seen, rivalling that of the Black-eared Bush-Tit of the Guatemalan highlands. The wall of this nest was very thick and soft. On the outer surface green foliose lichens (Jungermanniales) predominated. There was an admixture of green moss, and many fragments of greenish and grayish foliose and fruticose lichens. There were also many pieces of dicotyledonous leaves which had been brought green but had dried and become silvery on the lower surface. The largest of these pieces was slightly over an inch in length by half an inch in breadth. Conspicuous on the outside were also many bright yellow pieces of a closely woven, silky fabric, evidently the egg cases of a spider but possibly of some insect. There were likewise tufts of brown silk. All these varied materials produced the effect of a mosaic composed of innumerable pieces of different shapes and colors.

On the inside of the wall gray-green lichens predominated and liverworts were a minor component. A few slender grass blades were included in the inner part of the wall. The lining was composed largely of the same finely branched, gray beard-lichen (*Usnea*) of which the Long-tailed Silky-Flycatcher makes its nest. Mixed with the lichens were a few white horsehairs, a few fine straws, and some thin dry branches of a horsetail (*Equisetum* sp.). Cobweb had been abundantly used to bind together the amazingly heterogeneous components of this little nest.

Eggs and Incubation

In this nest in the Winter's bark tree of which I watched the construction, the first egg was laid early on 29 May and the second on the following day. These eggs, which I saw only reflected in the mirror that I held above the nest, were white, with small, scattered, dark spots on the broader end. Only two eggs were laid, and incubation began on the day that the second was deposited.

I chose 1 June as the day for watching the vireos incubate. It was not a fortunate choice, for it dawned with a steady wind driving clouds and drizzle over the mountain. I did not begin my vigil until 7:15, when sunshine was breaking through the cloud-mist, which had thinned enough to permit me to see the birds. Soon after I had settled in my raincoat on a rotting log near the nest, a huge motor truck came noisily up the road behind me, passing about 20 feet from the nest and stopping about 50 feet from it. Here four men began to load the truck with a number of large alder logs that had been piled at the top of the high bank on the opposite side of the road. The work proceeded with much talking, pounding, rattling of chains, and once a great, earth-shaking thud as several carelessly placed logs fell from the truck into the roadway. Then a yoke of oxen was brought to haul the fallen logs to the top of the bank, so that they could be replaced on the truck. All this bustle continued for over 3 hours, and more than once I was tempted to suspend observations; yet the vireos continued to incubate, as unperturbed as though they were in some secluded forest.

When I arrived at 7:15 a.m., a vireo was incubating in the drizzle, sitting with his face inward toward the nest tree. At 7:52 his mate came silently and replaced him. The change-over was effected rapidly, with no note audible to me, and the newcomer settled on the eggs facing inward. She remained until 8:33, when the male came singing, thereby revealing his sex and, indirectly, his mate's. About half a minute after her departure he settled in the nest, facing outward. For 50 minutes he sat in silence, then he sang loudly from the nest a few times. The female returned in silence to replace him after he had been present just an hour. She sat sideways, her head toward a supporting twig and her tail tilted strongly upward; unlike some antbirds and vireos of other species, these vireos might face in any direction while they incubated. After a while, when the gray cloud-mist was again drifting through the nest tree, the female turned to face outward. She stayed on the eggs for 58 minutes and left when she heard a male (her mate?) singing in the distance. The eggs then remained exposed to the chill mountain air for 27 minutes. At 10:58 a vireo approached the nest cautiously, making a circuit to reach it from the inside of the tree rather than the outside. It sat facing outward, and remained until relieved by its partner at 11:33. Since, after the interval of neglect, neither bird sang, I could not now recognize the sexes.

Before I could resume observations on this nest, the eggs vanished. Still, my noisy morning of watching left no doubt that both sexes incubated, taking sessions up to an hour in length, and it suggested that the male might be giving as much, or more, time to the eggs than the female, at least by day. In the vireo family there are numerous species in which both sexes incubate, and others in which only the female attends the eggs (Skutch, 1960).

The Nestlings

While watching a nest of the Long-tailed Silky-Flycatcher in late June, I so frequently saw Yellow-winged Vireos in the same alder tree that I suspected they had a nest there; but it was so excellently concealed amid the abundant foliage that it took me a long time to find it. There was scarcely any point on the ground whence this 25-foot-high nest could be viewed. The frequent arrival of a parent with food left no doubt that there were nestlings, which still required brooding;

but I could not reach the nest to learn how many were present. From 6:42 to 9:56 a.m. on 24 June, each parent brought food eight times. The only item that I could recognize was a larva in the female's bill. She always came to the nest in silence, whereas each time the male approached he sang his droll little song with food in his bill. If the nestlings were slow to take the food, the parent might utter a few very soft notes, and once the male sang a little while offering food to the young. All droppings were swallowed by the parents. After delivering food the female always brooded, a total of seven completed sessions, ranging from 5 to 24 minutes and totalling 100 minutes. The male brooded only five times, ranging from 7 to 16 minutes and totalling 60 minutes. The nestlings were left uncovered for four periods totalling 34 minutes. The female always sat facing the fork that supported the nest, her tail projecting over the outer side. The male usually sat transversely, his head over one supporting arm and his tail over the other. Once he sang a few notes while brooding, but usually he was silent, as the female always was.

Family PARULIDAE

FLAME-THROATED WARBLER

Vermivora gutturalis

The Flame-throated Warbler, which Ridgway called the Irazú Warbler, is about four and a half inches in length. As in many of the non-migratory wood warblers of the tropics, the sexes cannot with certainty be distinguished by their appearance. In both, the upper parts are plain slate-gray, with a large triangle of black on the back. The face is largely black. The chin, throat, and chest are on different individuals bright orange, flame-color, or yellow, and the more posterior under parts are white in the center, grayer on the sides. In some pairs, the male's throat is deeper in color than the female's.

This pretty warbler is confined to the high mountains of Costa Rica and adjacent parts of Panama, where it occurs on the Volcán Chiriquí. Although once I found an immature bird at an altitude of about 5,600 feet, the species is rarely seen below 6,000 feet, from which level it extends upward through the oak forests with dense undergrowth of bamboos to 10,000 feet or more. Although probably originally a forest-dweller, it has adapted itself to the pastures with scattered trees which now occupy so much of the more accessible portions of the high mountains of Costa Rica. In these pastures it is by far the most abundant permanently resident wood warbler, often the only one. It forages by preference well up in the trees, among the terminal foliage of the twigs, from which it gleans insects and many green larvae. At times it descends into low bushy growth on the sides of the deep ravines which intersect the mountain slopes.

Once in June, as the breeding season was ending, I found a flock of about 20 Flame-throated Warblers, mostly juveniles. Usually, however, these warblers are seen alone or in pairs, and this is especially true early in the year, as the nesting season approaches. Although they mostly avoid each other's company, Flame-throated Warblers may seek that of other kinds of birds. At the end of April, one foraged with a flock of migrating Black-throated Green Warblers. Once I watched a lone Flame-throated Warbler accompany a pair of Yellow-winged Vireos who were hunting in some spreading alder trees in a mountain pasture. It followed the vireos from tree to tree, gathering insects and larvae from the foliage much as they did, but flitting from branch to branch with a sprightliness that contrasted with their more deliberate movements. On another occasion, a pair of the warblers accompanied a pair of the vireos through the alder trees.

Voice

The Flame-throated Warbler's song is a weak, dry, rather harsh buzz, delivered with widely open bill. It resembles the song of the Crescent-chested or Hartlaub's Warbler of northern Central America but is even drier and more insect-like—if a 30-year-old memory and written descriptions may be trusted in making such a comparison. Strangely enough, in March and the first half of April I spent many hours watching two early nests of the Flame-throated Warbler without hearing a single song. Until the end of April, I heard these warblers sing only once, at a distance from any known nest. In May, when the earliest young were on the wing, the tuneless buzz sounded more freely from the scattered trees of the pastures, where chiefly I studied the Flame-throated Warbler. I often heard it in the middle of the day, and sometimes in the cold rain which now fell frequently. The period of song was brief, and by mid-June the buzz was rarely uttered. The only other note that I heard from this species was a very fine, weak *chip,* delivered by a female to whose nest I had climbed.

Nest Building

On 30 April 1950, I spent an hour watching a Flame-throated Warbler build her nest at an altitude of 9,500 feet on the southern side of Volcán Irazú. In 1963, I found five occupied nests between 6,500 and 7,500 feet above sea level at La Giralda. The first two of these nests were built in mid-March, when dry weather prevailed. The other three were built in April or May, after the wet season had begun; and from the latest the young departed at the end of June. Accordingly, the breeding season started toward the end of the dry season and continued into the early part of the long rainy season.

The six nests mentioned above are the only nests of the Flame-throated Warbler of which I have found records. They occupied sharply contrasting sites: three were well up in trees, three on grassy banks. Of the arboreal nests, two were in closely similar situations, on thick, lichen-covered, horizontal limbs, between tank bromeliads whose broad, strap-shaped leaves interlocked above them, forming a roof that seemed to be effective in shedding rain. One of these nests was 18 feet up above a roadway much used by pedestrians and cattle, amid pastures with scattered trees, at a distance from forest. The second nest was about 35 feet up in a tree of *Hedyosmum Artocarpus* growing in a pasture close by a large tract of heavy forest. The third arboreal nest was much higher, about 70 feet up near the top of a massive, epiphyte-burdened tree at the forest's edge. Here it was situated near the end of a thick ascending branch, amid moss, lichens, seed-bearing epiphytes of several species, and clustered, thin, upright, leafy shoots of the tree itself, which formed a canopy above it.

Of the terrestrial nests, the first was situated on a narrow shelf, about five feet above the base of a nearly vertical grassy bank beside a concrete highway. The nest site was well hidden behind a hanging tuft of grass. The other two nests were near the top of yard-high banks beside old, sunken cowpaths in the pastures at La Giralda. Each of

FIG. 5. Pasture with native alder trees (*Alnus acuminata*) and, in central background, introduced cypress trees (*Cupressus Benthamii*) at La Giralda, at about 7,000 feet, habitat of the Yellow-winged Vireo, Flame-throated Warbler, Flame-colored Tanager and Rufous-collared Sparrow, 18 June 1963.

these nests was set in a deep recess in the grass, moss, and other low vegetation that covered the vertical bank. One was particularly well concealed, as the narrow, inconspicuous entrance to its cranny was screened by a few small ferns.

Unlike the nests of many other tropical warblers, including species of *Myioborus, Basileuterus,* and *Ergaticus,* that of the Flame-throated Warbler is built without a roof; but all of the six that I saw were placed in situations where they were well protected from rain, either by dense pasture grass and moss or by epiphytes, particularly tank bromeliads with broad leaves. It is of interest that the related Crescent-chested Warbler likewise chooses both arboreal and terrestrial sites for its nest. The three nests of this species that I found in the Guatemalan highlands were all on the steep sides of deep, dry ditches, where they were well roofed by fallen dead leaves (Skutch, 1954:377-378). In Mexico, however, a nest of this species was built in a clump of ferns growing on the side of a mossy oak limb, 10 feet above the ground (Rowley, 1962:262). It was surprising to find a bird of another family, which shares the habitats of the Flame-throated Warbler in the Costa Rican highlands, selecting the same great range of nest sites. At La Giralda, I discovered nests of the Common Bush-Tanager sheltered beneath tank bromeliads high in trees and set amid grass on vertical banks. But in addition to these sites that could be matched by those of the Flame-throated Warbler, I have seen nests of the bush-tanager in situations rather different from any in which I have noticed the warbler nesting, such as amid the moss thickly covering the vertical trunk of a tree and on the ground on a steep hillside, beneath a bush.

At three nests that I watched while building was proceeding actively, I saw nothing to suggest that the male helped with the work. I found the nest in the *Hedyosmum* tree at 9:20 a. m. on 13 March 1963, and in the next 23 minutes the female brought material 26 times. Then she relaxed her effort, bringing only one additional billful between 9:43 and 10:00 a. m. On the following morning, I watched this nest from 6:05 to 10:05 a. m. In the first hour, material was brought 23 times; in the second hour, 21 times; in the third hour, 25 times; and in the fourth hour, 19 times—a total of 88 trips to the nest in four hours. As on the preceding morning, the material brought to the nest was chiefly green moss, plucked from neighboring trees, but sometimes gathered from limbs that had fallen to the ground. Not content with a single strand of moss, the warbler added piece to piece in her bill, and often flew to the nest with a conspicuous green bundle. More rarely she brought rootlets or pieces of leaf. Arriving with a contribution, she always alighted in the branches above the nest and hopped down to it. On some visits she deposited her load in the nest and left immediately, but on others she stayed to arrange the materials. Once she remained inside for 2½ minutes, and on five occasions she stayed between one and 2 minutes; but usually she was within for less than a minute. Long visits to the nest alternated with short ones; after depositing a few loads without giving much attention to their arrangement, she would on the next visit remain to shape her structure. From the ground, I could not see what she did inside the nest, other than that she moved around a good deal.

By 16 March, this warbler was bringing fine pieces for the nest's lining instead of moss. This material was gathered in the forest down the slope and carried into the nest so rapidly that only exceptionally could I distinguish it. Now the builder devoted much time to shaping her structure, remaining within once for 3¼ minutes and frequently for a minute or two. From 9:00 to 10:00 on this morning, she visited her nest 12 times.

While this female built, her mate, whose throat was of a slightly more intense flame-color than hers, was often in or near the nest tree, especially in the early morning. Frequently he pursued his partner as she flew toward or from the nest, and sometimes he seized her while she rested on her way to the nest with a laden bill. Twice the two fell, clutched together, from a high branch almost to the ground. Once the male hung from the female's tail in a manner reminiscent of the Green Honeycreeper (Skutch, 1962a:93, 96). As far as I could tell, he never visited the nest. Neither during building, incubation, nor while he helped to attend the nestlings, did I hear this male sing in the vicinity of the nest.

I found the highest nest, 70 feet up at the forest's edge, by seeing a warbler carry material to it at 8:12 a. m. on 17 March 1963. In the next hour, the supposed female visited it 27 times, bringing inconspicuous materials which were evidently for the lining. At no time during this hour was more than one bird in view, and I heard no song. At the nest that was being built on the grassy roadside bank on Volcán Irazú on 30 April 1950, the male was more attentive. While his duller mate built actively, he sometimes accompanied her to the nest, and he drove away trespassing birds.

Two nests, removed from niches in grassy banks after the young had flown, were great masses of green mosses and liverworts. One of them also contained a few filmy ferns and lichens in the foundation. Toward the interior, this mass of living cryptogamic plants was interlaced with fine fibrous materials, including light-colored vegetable fibers, rootlets, black fungal rhizomorphs, a few horsehairs, and the like; but even in the innermost lining of the shallow depression in the top of the mass, green moss predominated. One nest measured 3 inches in height by 4 to 5 inches in diameter. The cavity was 2¼ inches in diameter by 1¼ inches deep. Although the nest itself was an open, unroofed structure, it was, in each instance, covered above by epiphytic plants or the dense grass and moss amid which it was set on the side of a bank, as already mentioned.

Eggs

Of the three nests that could be reached, two contained sets of two eggs and the third held two nestlings. In one nest, the eggs were dull white, very faintly and finely sprinkled over the whole surface, but especially on the thicker end, with pale brown. In another nest, one of the white eggs was immaculate; the other was marked by a few, widely scattered, tiny, brownish spots. The eggs in the first of these sets measured 16.2 by 12.9 and 16.6 by 13.0 mm. Those in the other set were 19.0 by 13.1 and 18.0 by 12.9 mm.

At La Giralda in 1963, laying began (in accessible nests) in the second half of March, and the latest set was laid (to judge by the date of hatching) at the end of May.

Incubation

I passed the morning of 10 May watching the nest set between tank bromeliads on a horizontal branch 18 feet above a roadway. When I arrived at dawn, a cold wind was driving cloud-mist through the trees. By seven o'clock the wind had died away and the sun was shining, but by nine o'clock clouds covered the sky and remained for the rest of the forenoon. This chilly morning with little sunshine was rather typical of the weather at this season. At 5:19 a. m. the female left her eggs, which she had been incubating for a week. In six hours of the morning, she took eight sessions, ranging from 19 to 34 minutes and averaging 27.1 minutes. Her eight recesses varied from 10 to 22 minutes and averaged 17.0 minutes. She covered her eggs for 61 per cent of the morning. On each departure, she left her nest by a sudden drop that carried her almost to the ground before she broke her fall and flew away. Sometimes she skimmed low above the pasture for about 50 yards before she rose into a tree. The abrupt vertical fall from the nest is a mode of departure practiced by a number of small birds that breed in trees. It decreases the likelihood that they will betray their nest's position by leaving it; one usually does not notice them until they are well below the nest and their point of departure is no longer evident.

The male took no part in incubation, as far as I could learn. Although he frequently escorted his mate to the nest tree, he did not once go to the nest itself. Once when he approached it as she was returning to her eggs, she chased him away. Once he drove a trespassing Flame-throated Warbler from a neighboring tree. In the early morning, he sang as he accompanied his partner to the nest tree. After that, the only song that I heard came from somewhat distant trees, so that I could not tell whether it was delivered by the male of this pair or by some other Flame-throated Warbler.

At this nest, which was inspected periodically by raising a mirror attached to the end of a long bamboo pole and viewing the reflected image through field glasses, the two eggs were laid on 1 and 2 May. One had hatched by the late afternoon of 17 May, and at 6:30 a. m. on 18 May there were two nestlings. The incubation period was approximately 16 days, which is not unusually long for a tropical wood warbler (Skutch, 1954).

Young

The newly hatched nestlings have orange-pink skin and long but sparse gray down. The interior of the mouth, as revealed when they gape for food, is yellow on its marginal regions, including the inside of the upper mandible, but pink or pale red in its more central and deeper regions. This bicolored condition persists at least until the young are feathered. It contrasts with the self-colored mouth lining of most passerine nestlings, including those of warblers of the genera *Myioborus* and *Basileuterus,* whose gaping mouths are uniformly yellow or orange-yellow.

The nestling Flame-throated Warblers are fed by both parents, who bring them many billfuls of small green caterpillars. One day at the end of June, I looked into a nest in a bank very carefully, so as not to frighten the feathered nestlings; but one darted out and the other promptly followed. The first flew quite competently, covering about 50 feet on its first attempt to use its wings; but the second nestling fluttered only a few feet and then tried to hide in the pasture grass, where it was easily caught. As I took it in hand, both parents alighted on the ground with spread wings, but they scarcely fluttered over it, in a brief and unconvincing display of injury simulation.

The captured fledgling was well feathered. Its dorsal plumage was largely olive-gray, and in the middle of its back was a large blackish triangle, bordered behind by gray with no olive tinge. The wings and tail were dusky, and some of the upper wing coverts had narrow buffy tips. The lores and face were also dusky, and there was a narrow but distinct post-ocular stripe of buff—a mark not present on the adults. The chin and throat were orange-yellow, which faded to buff on the abdomen. When replaced in its nest, this young warbler, about 12 days old, remained. The other, evidently the older of the two, was about 13 days old and too alert to be captured for inspection. At an arboreal nest where the young were not disturbed, the nestling period was also approximately 13 days.

From one of the nests in a bank, the young departed leaving the front soiled with droppings; but another nest was left quite clean.

On the morning of 13 June, I found a flock of about 20 Flame-throated Warblers, mostly juveniles, in the trees along a sunken road between pastures. They kept together as they flew, a few at a time, from tree to tree. Several alighted on the bank beside the roadway, where at least two of the warblers picked up bits of moss or grass, then dropped them to fly on with the flock. This was the largest company of this rather solitary warbler that I ever met.

TROPICAL PARULA WARBLER

Parula pitiayumi

One of the smaller wood warblers, the Tropical Parula is scarcely four inches in length. The male's upper plumage is (in various races) blue-gray to blue, with a large contrasting triangle of olive-green in the center of the back. There are two (or in some races one) white bars on the wing coverts and more or less extensive white areas on the outer tail feathers. The lores and cheeks are black. The under plumage is bright yellow, more or less suffused with orange-tawny on the chest, and paling to white on the crissum and under tail coverts. The female is similar but duller. The black face with no white on the eyelids and the paler, less distinct pectoral band of the male serve to distinguish this species from the rather similar American Parula, with which it is sometimes considered conspecific.

One of the most widely distributed of the resident warblers of the tropics, the Tropical Parula breeds from southern Texas through Mexico, Central America, and South America to Peru, Bolivia, northern Argentina, and Uruguay. Although not only at the extratropical limits of its range but even well within the tropics this warbler is found near sea level, in many parts it is confined to greater elevations. In Costa Rica, it occurs from about 1,000 to 6,000 feet on the Caribbean slope, while on the drier Pacific slope of the southern part of the country it is rarely seen below 3,000 feet. In the Santa Marta region of Colombia, Todd and Carriker (1922:454) found the species from 1,000 to 4,000 feet or more, rarely lower. In Venezuela, however, its vertical range extends from near sea level up to 8,000 feet (Phelps and Phelps, Jr., 1963:321), and in Trinidad it inhabits the better-wooded parts of the island at all elevations (Belcher and Smooker, 1937:520).

In various parts of its far-flung range, such as southern Texas and eastern Venezuela, this warbler frequents low, dry woodlands, and near Cali in the Cauca Valley of Colombia I met it in semi-arid cultivated valleys with scattered trees. But in the mountains of Costa Rica, as on the eastern slopes of the Andes in Ecuador, I have always found it in lofty, humid forests, where it forages well above the ground amid the foliage of the great trees and the masses of epiphytes that they uphold. Usually I have seen these warblers at the forest's edge, but this may well be because of the difficulty of glimpsing such small, high-ranging birds while standing in unbroken woodland rather than because they prefer its margin to its interior. Possibly because of difficulties of observation, I lack evidence that Tropical Parulas remain mated throughout the year, as many other tropical warblers do.

Like other wood warblers, the Tropical Parula subsists largely on adult and larval insects that it captures while flitting tirelessly through the foliage, but it favors at least one item of vegetable food. At higher altitudes, the guarumo trees (*Cecropia* spp.) frequently escape infestation by the Azteca ants which in the lowlands commonly establish populous colonies in the wide central hollow of their trunks and branches. On these antless trees there is an accumulation of the ants' staple food, the tiny white protein corpuscles that are produced on the brown, hairy cushions at the bases of the long, stout petioles of the ample palmate leaves. I have often watched Tropical Parulas cling to these petioles while with their sharp bills they gathered many of these dainty tidbits, to eat or to carry to their young. Other small birds, including Red-faced Spinetails, Bananaquits, and Yellow-thighed Finches, compete with the Tropical Parulas for the same diminutive but doubtless very nourishing particles.

Voice

The Tropical Parula's song is a rapid flow of small, bright notes which often run into a fine, high-pitched trill. Sometimes the song's concluding notes are again more widely spaced. For persistence of delivery the Tropical Parula has few equals, even among the songful wood warblers. Not only does it sing profusely in nesting time; it continues through most of the year, even at seasons when most of its neighbors are silent. On the northern slope of the Cordillera Central, it sang freely in August and September, when the great majority of birds of all kinds had finished breeding; and although I heard it less often in the dark and stormy months which followed until the year's end, it was still one of the most persistent songsters. Likewise in January on the Pacific slope of southern Costa Rica, when days full of brilliant sunshine depress the singing of many birds accustomed to cloudier weather, the Tropical Parula is one of the most tuneful birds of the forest. And as it sings through the year, so also it continues through the day, often performing profusely during the midday lull in avian activities.

NESTING

On 6 May 1964, I saw a female parula, with material in her bill, fly into a thick cushion of green moss surrounding a nearly upright branch about 2 inches thick, 30 feet up in a slender tree. This tree stood in a light open grove in a bushy pasture, close by a tract of heavy forest, at an altitude of 4,000 feet above sea level near Las Cruces, in extreme southern Costa Rica. The cushion of moss into which the warbler vanished, through an inconspicuous round opening in its side, was fringed with some dangling strands of moss; but the nest was being built in the more solid portion of the green mass, evidently against the lower side of the inclined branch that it enveloped.

The warbler built at a leisurely pace. In the first hour after I found the nest (7:55-8:55 a. m.), she brought material only 7 times; and in the following hour (8:55-9:55), eight times. Early next morning she was working faster, making eight trips to the nest in the half-hour from 6:35 to 7:05. As far as I could see, only the female built, and she brought only fine fibrous materials to line her niche in the mossy cushion. From time to time her mate sang brightly in the surrounding trees.

I could not reach the nest to examine the eggs; nor, from the ground, could I see the female incubating in the midst of the moss. But on 5 June, when I returned to this nest after an absence of three weeks, I could see three well-feathered nestlings in the doorway, whenever they stretched forth their open mouths for food. In the hour from 11:30 a. m. to 12:30 p. m., when the sky was lightly overcast, both parents brought them food 52 times, or nearly once every minute. Among the articles I recognized were green caterpillars, a small, green, grasshopper-like insect, and once, apparently, tiny white protein corpuscles from the Cecropia tree, which the parent had difficulty transferring to the young. Many items were too small to be recognized as they were rapidly passed to the nestlings. Droppings were carried away in the parent's bill; if one of the white sacs started to fall, the parent darted down and caught it in the air. The nestlings made a rapid clicking sound when food was brought to them. Their busy father sang sparingly.

This was the only nest of the Tropical Parula that I could find. In other parts of its extensive range, the species chooses diverse sites. Two nests found in Trinidad by Belcher and Smooker (1937:521) were each placed in a growth of cactus hanging beneath a horizontal branch of a ceiba tree, at about 30 feet above the ground on a steep hillside. Built of green moss, these nests were roofed structures with a side entrance. Each contained two eggs, one in early June and the other in early July. On an almost glossless white ground, the eggs were spotted and blotched with deep or pale chestnut, forming a more or less distinct cap on the broader end.

A nest found near Brownsville, Texas, was hollowed out of a thin bunch of hanging moss, and apparently building was confined to carrying in a few horsehairs as lining. This nest, which also had a lateral entrance, hung only 10 feet from the ground. Other nests in Texas and northern Mexico were in hanging masses of *Tillandsia,* or of "a gray mistletoe-like orchid . . . very common on the Rio Grande." These nests were also chambers which the warbler had evidently formed by separating the massed stems of the pendent epiphytes, but they were more abundantly lined, with short cottonwood fibers in the case of the nest in the orchid plant, and with black and white cattle hair in the *Tillandsia.* A nest that was "dug and hollowed in a bunch of pipestem mosses" was still more elaborately made. It was a compact little structure of fine rootlets and grasses, shreds of the brown inner bark of the palmetto, and some weed blossoms; with finer shreds, a little plant down, and a few feathers for the inner lining (Bent, 1953: 150-151). These nests of the northernmost race of the species contained three or four eggs, not unlike those described from Trinidad.

CHIRIQUI YELLOWTHROAT

***Geothlypis chiriquensis*[1]**

About five inches long, the male Chiriquí Yellowthroat wears the mark that distinguishes his genus, a broad band of black which extends from his forehead over his face to the ear coverts. This conspicuous black mask is bordered along its upper margin, from the crown to above the ear coverts, with gray. The remaining upper plumage is plain olive-green, with no wing bars. The under parts are yellow, very bright on the throat, but clouded with olive on the flanks. As in other yellowthroats, the female is more plainly attired, with olive-green upper plumage which becomes gray on top of the head. There is an indistinct light superciliary streak above the pale gray of her orbital region and cheeks. Her throat, breast, and under tail coverts are bright

[1] Often considered conspecific with *G. aequinoctialis* of South America. — ed.

yellow, which becomes paler on the abdomen and is clouded with olive on the flanks. In both sexes, the bill is black, the legs and toes flesh-color.

This warbler, related to the Masked Yellowthroat of South America, was long known to occur only near the base of Chiriquí Volcano in extreme western Panama. Across the border in Costa Rica it was not recorded until 1964, when I found it well established in the region of Cañas Gordas and San Vito de Java, between 3,500 and 4,000 feet above sea level on the Pacific slope. Here it was not beyond sight of the imposing volcano whose name it bears, which, however, is locally known as Volcán Barú. These yellowthroats inhabited well-drained grassy and weedy areas where scattered trees or shrubs stood above lush herbage. In the low, dense cover that they preferred, they were difficult to see, except when the male rose to a bush or the low branch of an isolated tree to sing, as he often did in May and June.

Voice

The Chiriquí Yellowthroat is as fine a songster as the Olive-crowned Yellowthroat of the Caribbean slope of Central America, and this is high praise. The song which the male pours forth from a perch, at no great height above the herbage where he habitually lurks, is among the most beautiful and stirring performances that I have heard from wood warblers. It is a rapid flow of varied notes, beginning low and strong, becoming higher and weaker as it gathers speed, and often ending in a little garbled flourish. One yellowthroat seemed to sing *Tuwitchywer tuwitchywer tuwitchywer tuweecha tuweecha tuweecha che che che chit.* But the yellowthroat's delightful song is by no means stereotyped. This individual sang a number of versions, all similar to the one that I have attempted to paraphrase; other individuals used quite different phrasing, although their songs were of the same type. The final rapid, high-pitched flourish (which is sometimes omitted) is even more difficult than the opening notes to represent by written syllables. Sometimes the song is much longer than the one that I have tried to transcribe, or perhaps a number of songs are run together with hardly a break between them; and since an average song lasts about 5 seconds, these compound songs continue for a good fraction of a minute. On rare occasions this yellowthroat sings in flight, as does the Olive-crowned Yellowthroat. In June I watched a Chiriquí Yellow-throat rise steeply to a height of 15 or 20 feet above a weedy pasture, singing loudly, then drop as steeply downward to vanish amid the weeds. Another male sang in an undertone while describing a lower arch in the air.

A building female sometimes uttered a fine, low *chip*. From time to time she opened her mouth widely, exposing its flesh-colored interior, and with her mandibles hardly moving gave a long drawn, sharp rattle or *churr*. Sometimes this seemed to be merely a call to her mate, but at other times it appeared to express anxiety or mild alarm. Once she rapidly repeated whining notes which, to judge by the posture she assumed and her mate's prompt response, were an invitation to him.

Nest Building

On 21 May 1964, I found a female working at a half finished nest near Cañas Gordas. The nest was 18 inches up in a dense clump of lemon grass about 3 feet high. This was one of a number of similar clumps growing along a barbed-wire fence, between a weedy pasture and the grassy dooryard of a deserted house. When I returned at 6:48 the following morning, when a brilliant sun stood just above the distant Chiriquí Volcano in a clear blue sky, the female warbler was building actively. She was gathering broad dry blades from clumps of lemon grass several yards from the nest. Usually she carried only one blade at a time, especially if it was long, but occasionally she took two in her bill. On her way to the nest she rested one or more times on the intervening strands of wire—up to about half a dozen times if the blade in her bill was unusually large. Then she flitted to some young sprouts of a bush that grew amid the grass, and from this she reversed her course to reach the clump where her nest was hidden.

While I sat unconcealed beneath a tree about 50 feet away, the female warbler, seeming to pay no attention to me, proceeded with her task. In the first 34 minutes after my arrival, she took 28 billfuls of material to the nest. Then she stayed out of sight for 11 minutes. At 7:33 she returned with a billful of fibers, the first material of this sort that I saw her bring, but on following trips to the nest she carried more grass blades. By 7:48, an hour after my arrival, she had brought 34 billfuls to the nest. After this initial spurt of activity her effort relaxed, possibly because she was perturbed by a hen who wandered around the abandoned house, or by a cat that I briefly glimpsed. In the second hour (7:48-8:48), she brought only four contributions to the nest.

On the next day, 23 May, I arrived long before the yellow-throat began to work. She resumed building at 5:51 a. m., as the sun floated up into a clear sky above the distant volcano. By 6:51 she had brought 19 contributions, and in the next hour she added 12 billfuls to her nest. On nearly every visit she now brought fine fibrous materials for the lining, which she gathered in a neighboring bushy hedgerow, or beyond it. In passing back and forth, she flew low above the weedy pasture. Arriving with a sheaf of fibers, she alighted on the wire fence, and might rest on it several times more as she proceeded along it to the bush from which she went to the nest. This detour needlessly increased the length of her journey, and before long these rests on the wire were omitted. Although on the preceding morning her mate came near the nest only once, today he was more attentive, often following her across the pasture and alighting on the fence near the nest, which on one occasion he seemed to visit. (It was hidden from my view.) He sang more than on the first morning that I watched, but still sparingly. Once he sang in flight.

By noon of 24 May, the third day after I found the half-finished nest, it appeared finished. The bulky open cup was composed largely of flat blades, up to half an inch in breadth, of the lemon grass, which stood up untidily around its rim. In the outer layers of the nest only such blades were used. The lining was composed chiefly of long, round, smooth, brown fibers of unknown origin. The external dimensions of the cup were 4 by 5½ inches in diameter by 3½ inches high. The interior was 2½ by 3 inches in diameter by 1½ inches deep.

Eggs and Nestlings

Although the nest appeared finished by noon on 24 May, a little more lining was added on the following day; and on 26 May the first egg was laid. With the laying of the second egg on the following day, the set of two was completed. These eggs were white, spotted with deep brown, pale brown, and pale lilac. The spots were largest and most crowded on the thick end of the egg and diminished toward the opposite pole, where they were few, small, and faint. The eggs measured 20.0 by 13.9 and 19.6 by 14.1 mm. The shells had little gloss.

I did not study incubation at this nest, because to make it visible would have necessitated cutting away much of the grass which surrounded and protected it. On a number of visits, I saw only the female sitting. Whenever I came in view, she jumped from the nest and vanished beneath the tall lemon grass, to reappear, after an interval, a good distance away. Both eggs were pipped at 11:00 a. m. and also at 4:30 p. m. on 10 June, and by 7:00 next morning they had hatched, after 14.5 to 15 days of incubation. The shells had already been removed.

The nestlings' skin was flesh-color, with a tinge of orange. Their natal down was pale gray. The bill was pale yellow, with flanges of a deeper yellow color at the corners of the mouth. The surface exposed when the mouth was opened was yellow on its outer portions (the inner surfaces of the two mandibles and along its margin), but pale red on its deeper portions, as on the tongue. The legs and toes were yellowish.

On the morning of 14 June the three-day-old nestlings were vigorous and had sprouting pinfeathers. That afternoon rain fell long and hard. Next day the two young yellowthroats lay dead in their nest. Evidently the parents could not adequately attend them through the rainy afternoon. The breeding season was waning, and I looked in vain for a replacement nest.

GOLDEN-CROWNED WARBLER

Basileuterus culicivorus

The golden spot for which this small warbler is named is not conspicuous; it occupies only part of a narrow yellowish stripe that passes from the forehead over the crown and hindhead. This median stripe is bordered laterally by bands of black, which in turn are margined by superciliary lines of dull yellow. An indistinct dusky streak passes through each eye. The dorsal plumage, posterior to the head, is plain greenish or grayish olive, and there are no wing bars. The cheeks are grayish, and the entire under parts from chin to under tail coverts are yellow, washed with olive on the sides and flanks. As in other species of this genus, the sexes are similar in appearance; but in one pair I learned to distinguish them by the color of their legs and toes. Those of the male were light orange, or between orange and flesh-color, while those of the female were pale flesh-color. The male's upper mandible was dusky, his lower mandible yellowish horn-color. The eyes of both sexes were dark.

The Golden-crowned Warbler ranges widely across the American tropics from northeastern Mexico (whence it has been known to stray into the Río Grande delta in Texas) to Uruguay and Argentina. In Middle America, it is confined to lower middle altitudes, from about 1,000 to 4,500 feet in Mexico and Guatemala. In Costa Rica, its over-all range has been given as 600 to 7,000 feet (Carriker, 1910:795; Slud, 1964:334), but it is certainly rare or local at both extremes. On the excessively wet northern slopes of the Cordillera Central, I found it at 2,500 feet near the Sarapiquí Lakes, but I failed to meet it once in the course of a year's residence at 5,500 feet below Vara Blanca. On the Pacific slopes of the Cordillera de Talamanca, where the dry season is more pronounced, this warbler remains higher, hardly dropping below 3,000 feet in the basin of El General. At 5,000 feet on the Río Cotón near the Panamanian border in February, it seemed to be the most abundant bird in the undergrowth of the heavy forest. It was also very numerous around 4,000 feet in the neighborhood of Cañas Gordas.

Golden-crowned Warblers are active birds, moving through the dark undergrowth of high forest so restlessly that one rarely enjoys more than a fleeting glimpse of them. At least in southern Costa Rica, they seldom venture beyond the dense humid woodland into adjoining areas of lighter vegetation. Mostly they search for their insect food in the layer of shrubs and saplings within 15 or 20 feet of the ground, yet they are rarely seen on the ground except when nesting. Nearly always they go in pairs, which often keep company with other small birds of the forest undergrowth, such as the Slaty Antwren, Spotted

Barbtail, Red Ant-Tanager, Sulphur-rumped Myiobius and other sylvan flycatchers. The slight rapid ticking, which these warblers keep up with scarcely an intermission as they flit through the undergrowth, calls attention to them and their small companions.

The male Golden-crowned Warbler's song, consisting usually of five slight musical notes sliding rapidly upward, is enchanting in its delicate simplicity. It reminds one of balmy days in early springtime, butterflies flitting through the sunshine, milkweed plumes or thistledown floating on soft breezes—things untouched by heaviness and care.

Nest and Eggs

Despite much searching through tall mossy forests where Golden-crowned Warblers were among the most abundant birds of the undergrowth, I have found only one of their nests. I should have passed right over this nest without noticing its presence, or perhaps crushed it beneath a foot, if the solicitous parent had not darted out of the ground litter just ahead of me as I ascended a gentle slope through fairly open forest. Her long-continued, realistic "injury feigning" assured me that she had eggs or young close by; nevertheless, I probably would not have found her nest if I had not noticed some green moss in the forepart of its roof, contrasting with the brown fallen leaves and twigs beneath which it was excellently hidden. The low, wide doorway of the oven-shaped structure faced down the slope and was shaded by the green leaves of a small plant of *Carludovica* that grew close by.

Some weeks later, after the young had left, I removed the nest from the little depression in the ground in which it had been set beneath fallen leaves. It was a roughly globular structure about 5 inches in diameter, with a round opening in its exposed side. The outer wall was composed largely of black rootlets but there was a variety of minor ingredients, including vegetable fibers, narrow decaying strips from the leaves of palms or other monocotyledons, and fine branching stems of foliose liverworts, with a few tiny green leaves remaining on the ends of some. The thick inner cup that held the eggs was composed of very fine, brownish fibers, densely felted.

This nest was situated on Finca "Loma Linda" near Cañas Gordas, at an altitude of about 3,700 feet. When found on 15 April 1964, it held three eggs, which were broadly ovate, white with a heavy wreath of dark brown blotches around the broadest part, and a scattering of paler brown spots over the remaining surface, especially the thicker end. These eggs measured 17.9 by 14.1, 17.8 by 13.8, and 18.7 by 14.9 mm.

On the island of Trinidad, another race of the Golden-crowned Warbler builds its nests either among fallen dead leaves on a gentle slope or, especially at higher altitudes, in a niche on a moss-overgrown cut bank beside a path. The domed structure with a wide lateral entrance is composed of grass and other soft vegetable materials and lined with fine grass. Three or four eggs are laid, the larger sets being usual at the higher altitudes. These eggs are white, marked with two shades of reddish brown; the measurements of six average 17.5 by 13.6 mm. In contrast to the situation in Central America, on Trinidad, as on neighboring islets and likewise the nearby mainland in Venezuela, the Golden-crowned Warbler resides from sea level up into the Subtropical Zone (Belcher and Smooker, 1937:522-523; Phelps and Phelps, Jr., 1963:337).

Incubation

In the afternoon of the day after I discovered the nest, I set my blind before it, beneath the lacy fronds of a tree fern. When I entered the blind as it grew light in the forest at 5:40 on 17 April, many birds were singing, including Great Tinamous, Bright-rumped Attilas, Lowland Wood-Wrens, and White-throated Thrushes. Migrating Olive-backed Thrushes were calling in the undergrowth around me. A pair of Orange-billed Sparrows hopped over the ground past the blind. Presently I heard a Golden-crowned Warbler's modest refrain from the forest down the slope. In the dim light I could not see whether the female warbler was in her nest. Evidently she was not, for at 6:03 she returned alone and entered.

After incubating her three eggs for half an hour, the warbler left the nest in an unexpected fashion. In front of her doorway a long thin stick leaned against a slender, upright, dead stem about 7 feet high. The warbler walked up the inclined stick to the stem, then climbed up the stem to its top, from which she flew off through the forest. On two subsequent departures, she chose the same route. At other times she left her eggs to creep unobtrusively through the low herbage beside the nest until she reached the base of a neighboring sapling, up which she climbed and flitted to the top of the vertical stem. On one occasion, she ascended in this manner to a height of about 11 feet before she flew. In this way she rose unobtrusively to that level of the forest in which she chiefly foraged, before she went off in search of food and doubtless also her mate. Once she returned to the nest by hopping down the inclined stick up which she sometimes climbed on leaving, but more often she reached her doorway by flitting through the ground vegetation. This last stage of her return journey was performed in discreet silence, although often she uttered ticking notes as she approached through the forest.

The warbler sat on her eggs with her head at either the right or left side of the doorway. From time to time during a session of incubation she shifted from one position to the other, but she never faced straight forward. After about seven o'clock there was enough light beneath the tall trees to see her head in the doorway, but her body was difficult to distinguish except when a spot of sunshine fell directly on the nest. By noon she had taken five sessions, ranging from 31 to 52 minutes in length and averaging 44.2 minutes. Her five recesses ranged from 22 to 32 minutes and averaged 26.8 minutes. She covered her eggs for 62 per cent of the forenoon.

The male warbler remained out of sight all morning, but until after nine o'clock I sometimes heard in the distance a song that was evidently his.

Nestlings

When I visited the nest early in the afternoon of 19 April, the parent hopped out to creep slowly over the ground for about ten feet. After giving this distraction display, she climbed up a slender stem, from which she took flight. The three eggs had hatched since the preceding day. The sightless nestlings had pink skin with sparse gray down. The inside of each mouth was yellow.

When these nestlings were three days old, I watched them for the first 6 hours of the morning, which was mostly cloudy. Before the morning was over, I learned to distinguish the male parent from the female by the deeper color of the legs and toes of the former. Together they fed the three nestlings 26 times, of which I credited 15 to the female but only 7 to the male, who stayed away for long periods. At four feedings I could not identify the parent. At his first feeding, in the dim light before sunrise, the male whispered a few songful notes as he approached the nest with food in his bill. Thereafter, he always came in silence, as his mate did. On two occasions the parents delivered food while standing side by side in the wide, low doorway, their tails outward. As far as I could see, the nestlings' meals consisted wholly of mature and larval insects, but the parents usually approached the nest so rapidly that it was difficult to distinguish what they carried. Sometimes they brought what looked like an overflowing billful of tiny fragments of leaves. The parents usually flew directly to the nest, but occasionally they alighted on the inclined stick in front of it and hopped down along this route to ground level.

The male never stayed to brood the nestlings, but after feeding them the female usually settled down to warm them. Her 13 sessions of brooding ranged from 3 to 13 minutes and averaged 8 minutes. She was in the nest only 30 per cent of the forenoon, already a considerable reduction from the time she spent sitting while she incubated. If her mate arrived while she was brooding, she always left, so that he could feed the nestlings directly; she never took food from him. After a spell of brooding, the female always flew directly from the nest and continued until out of sight, in contrast to her slow, scandent departures during the period of incubation. Whenever I could identify the parent who now left by climbing up the stick and the sapling against which it leaned, it was the male. When beyond sight he sometimes sang a little. The parents usually swallowed the nestlings' fecal sacs, but sometimes they carried them away in their bills.

On 26 April, when seven days old, the nestlings were in pinfeathers, and two days later they were clothed in plumage, at least on the back. On 29 April I again watched them through the first 6 hours of the day. They were fed a total of 82 times, or at the rate of 4.6 times per nestling per hour. They received only seven meals between 5:25 and 6:25, but from 9:25 to 10:25 they were fed 19 times, the maximum number of meals in a single hour. Often the parents approached so swiftly with food that I could not distinguish them; but I was sure that both were attending their offspring, for sometimes they delivered food while standing side by side. The rapid delivery of the food made it difficult to see, but again I noticed only insects in the parents' bills. There was no brooding on this clear morning. All the nestlings' droppings were now carried away in the parents' bills and none was swallowed. When carrying off a dropping, the parent always flew directly from the doorway, but when leaving with empty bill it sometimes chose the route of the inclined stick and the sapling.

The bright-eyed nestlings had become rather noisy, twittering loudly whenever they were fed. If they were hungry and noticed a parent's approach, they began to call even before it arrived; and often they continued for some seconds after it had fed them and departed. Occasionally a nestling uttered a little *chip*. At the end of my watch, I cautiously approached their nest for a better look at them; but without being touched, they suddenly "exploded" before I could see much. The ten-day-old young could fly only a few feet, but they hopped and flitted rapidly away over the ground, chipping loudly as they went. They rested on perches only a few inches above the ground. They were well feathered, olive above and buffy below, but their tails were very stubby and their heads, on which the feathers were still mostly ensheathed, lacked the conspicuous stripes worn by their parents, who led them down the hillside through the undergrowth.

BLACK-CHEEKED WARBLER

Basileuterus melanogenys

The Black-cheeked Warbler has a bright chestnut crown, margined on each side by a narrow stripe of black, which meets the opposite stripe on the forehead. Each black band is in turn bordered by a conspicuous white superciliary stripe, below which is the black cheek. The upper parts of the body are largely olive. The tail and wings are more grayish, and the latter lack conspicuous bars. The under parts are varying shades of dull white and pale gray. The upper mandible is black with horn-colored margins and tip, and the lower mandible is yellowish horn-color. The legs are flesh-color clouded with dusky. The sexes of mated pairs are indistinguishable by their appearance, and both are about five inches in length.

This attractive warbler is confined to the highlands of Costa Rica and neighboring parts of Panama. In the latter country, it has been recorded as low as 4,500 feet (Ridgway, 1902:752), but in Costa Rica it appears to remain higher. Here the lowest point at which I have seen it was about 5,300 feet above sea level on the northern slope of the Cordillera Central below Vara Blanca, where it was rare; and Carriker (1910:797) found it from "about 6,000 feet nearly if not quite up to timber line on the high volcanoes." Between 9,000 and 10,000 feet it appears to be fairly abundant on most of the high mountains of Costa Rica.

A quiet, self-effacing bird, the Black-cheeked Warbler inhabits the undergrowth of the mossy cloud-forest on exposed summits, the bushy growth that chokes ravines, and the dense thickets of cane-like bamboo beneath the great oaks and other trees of the mountain forests. Through such tangled vegetation it flits restlessly, easily eluding the bird-watcher who tries to follow, as it hunts insects near the ground. Unlike many other birds of the mountain forests, it rarely ventures forth to forage in adjacent clearings and pastures, where its acquaintance could more easily be cultivated. It appears to travel in pairs, even in November when its breeding season is months away; and I have seen nothing to confirm the statement of Carriker (*loc. cit.*) that these "birds go about in small bands of from six to ten or more." A single pair of Black-cheeked Warblers often accompanies a mixed flock of woodland birds, including such species as the Highland Wood-Wren, Chestnut-capped Brush-Finch, Collared Redstart, Common Bush-Tanager, and Spotted Barbtail—never, in my experience, all of these birds together, and usually only one or two adults of a species in a single mixed flock.

Voice

The male Black-cheeked Warbler has a slight, lisping song, in form reminiscent of that of the Chestnut-capped Warbler, but much weaker. I have heard this song but seldom. A male whose mate was building very actively, early in the nesting season, sang most sparingly. Another male, whose mate hesitated long to return to her eggs in front of a newly erected blind, frequently repeated his shrinking refrain; but at other times I heard little song from him, too. The call of this warbler is a weak *chip*.

Nest Building

On the morning of 21 March 1963, I found a Black-cheeked Warbler starting a nest in a ravine near Los Cartagos in the Province of Heredia, at an altitude of about 6,300 feet above sea level. The ravine had a rather wide, flat bottom, through which a rivulet flowed. Its floor and sides were covered with small trees, shrubs, and herbage, forming a tangled growth intersected by cowpaths on the flatter areas. The nest's site was a shallow niche near the foot of a steeply sloping side of the ravine, where it was well concealed by fallen leaves and low herbage, and deeply shaded by the shrubbery and trees.

While I stood unconcealed about 20 feet in front of the nest, the female warbler, seeming not to notice me, built with a concentrated activity that I cannot recall having seen surpassed by any other bird. It was about 8:20 a. m. when I found the nest and began to count the builder's visits to it. In the next 15 minutes she brought 29 billfuls. Then she went off and stayed out of sight for 17 minutes. Returning at 8:52, she threw herself into her task with such amazing energy that in the next hour she brought material to her nest 102 times. Her interval of most rapid building was from 8:52 to 9:20, when she made 54 trips to her nest, or at the rate of nearly two per minute. After 10:20 she brought only one more billful, then went away. From 8:20 to 9:20 she had made 83 trips to nest, and in the next hour 71 trips—a total of 154 visits in the two hours from 8:20 to 10:20 a. m.

Most of the building materials were gathered from the ground or the bases of the plants within a few yards of the nest. At first the warbler brought chiefly rootlets, fragments of decaying leaves, and similar things. Soon she began to include light-colored, dry bamboo leaves as long as herself. These large leaves were usually brought singly; but when finer materials were selected, a number of pieces were carried together in her bill. Sometimes the warbler picked up her material a few yards from my feet. Usually she stayed in the nest just long enough to deposit her contribution, but occasionally she remained longer and arranged the materials with vigorous movements of her feet and bill. Sometimes the leaves which she loosely deposited screened the entrance to the nest, which promised to be a domed structure, making it difficult to distinguish from the surrounding fallen leaves, and confusing the warbler on her next visit: she might go to the bank a few inches from the nest, and sometimes she deposited her load there. At times she was hard to see, even from my post directly in front of the future doorway, after she had pushed into the nest through the loose screen of leaves. Presently with her bill the bird would push the obstructing leaves sideward and upward, clearing her doorway. At the end of the morning, I found the ground at the bottom and back of the niche with scarcely any covering. The builder was pushing all her materials to the top and the forward part of the side walls, leaving the bottom to be covered last, as seems usual with nests of this type. The 154 billfuls that I had watched her bring did not go far: her nest was still barely begun.

Most of the time the female warbler worked in silence. Occasionally she uttered a sharp *chip,* and rarely a series of these slight notes, which might have been a rudimentary song.

During the first spell of building that I watched, the male flitted through the bushes around the nest, but he did no work, as far as I could see. Just before the female began her second spell, he came, sang a few times, and entered the nest once or twice; but I was not sure that he brought anything to it. Then he vanished, and for the next hour and a half his indefatigable partner labored alone.

At dawn next morning, I hurried far down the mountainside to continue observations on this nest. A strong easterly wind was bearing ashes from Irazú, the huge, sprawling volcano 20 miles away, which a few days earlier had started an eruption that was to continue for the next year and a half. As I sat watching the nest, fine cinders sifted down into the sheltered dell, covering me and all the surrounding foliage with a thin layer of gritty dust. An hour after daybreak, the light was still so dim, the air so full of falling ash, that I was becoming alarmed. But by 7:30 the ash-fall had diminished and dusty sunshine slanted into the ravine, but the warblers failed to appear. For the next three days, cold, hard winds continued to blow. Even after calmer, milder weather returned on 25 March, work on the nest was never resumed.

Despite much searching, I found no other nest of the Black-cheeked Warbler until 3 June, when I discovered one in rather open forest with a dense undergrowth of bamboo, far up the mountain at an altitude of about 7,500 feet. This nest, in which incubation was already in progress, was situated on a low, nearly vertical, mossy bank, beside a little-used cowpath. Two feet above the base of the bank, it was supported by the low vegetation that sprang from the earthen wall rather than set in a niche. The bulky, oven-shaped structure had a substantial roof and a wide round doorway in the outer side. The foundation was composed of a mixture of long, slender bamboo blades and shorter, broader dicotyledonous leaves. The sides and roof were made largely of light-colored dead bamboo leaves, with which were mixed many pieces of fern fronds, chiefly of small polypodies and filmy ferns, and long, slender roots and rhizomes, which seemed to come mostly from epiphytic ferns. Toward the interior of the nest the material became finer; and the inner cup, which held the eggs, was thickly lined with shredded vegetable fibers and an admixture of brown scales from the fronds of large ferns. This nest was about 7 inches in height, 5½ inches from side to side, and 4½ inches from front to back. The interior measured about 2½ by 2 inches in diameter by 3½ inches in height.

Eggs

This nest, the only completed structure of the Black-cheeked Warbler of which I can find a record, contained two eggs on 3 June. They were short-ovate, glossy, white, flecked and speckled with cinnamon-rufous, the pigment deep and concentrated on the broader end, pale and thinly scattered over the remaining surface. Both eggs measured 19.0 by 15.0 millimeters.

This late set of eggs may have been a second brood but was more probably the replacement of a lost brood. On 26 May, I had found parents attending a stubby-tailed fledgling that could fly only a few feet, in a deep ravine overgrown with tall bamboo.

Incubation

The incubating female sat so steadfastly that I could look into the nest with my eyes only about 8 inches from hers. Doubtless I might have caught her by clapping my hand over the doorway. When I advanced my hand slowly, she jumped out while it was still a few inches away, dropped into the pathway at my feet, and crept over the ground with relaxed, quivering wings, continuing this excellent distraction display until she had passed from view amid the bamboos down the slope. She gave such displays repeatedly, both before and after her eggs hatched.

Although the female warbler permitted such a close approach if I found her already sitting, she refused to return to her eggs in my presence, even when I rested quietly at a good distance. Accordingly, to watch her incubate, I was obliged to use a blind. To avoid the risk of damage to the cloth structure by the cattle which roamed through this part of the forest, I did not, as is my practice, set up the blind at least a day before I intended to watch the nest; but as soon as it was ready, I began my vigil. It was then 6:37 on a dark and gloomy morning, when the forest around me was dripping from the shower that fell as the night ended. Presently the female returned, to flit through the bamboos and bushes behind and above the nest, advancing toward it only to retreat again, over and over. Her oft-repeated weak *chip* drew her mate, who flitted around with her and accompanied her on her uncompleted approaches to the nest, He left her, but was drawn back by her persistent complaints, again and again. He repeated his slight, lisping song more often than I have heard a Black-cheeked Warbler sing on any other occasion. After 35 minutes of hesitancy, the female returned to her eggs, which she incubated for a half-hour. Her next return to her nest was a repetition of the previous one; but before long she accepted the blind and with little hesitation went to her nest at the end of each recess. Her mate now paid less attention to her, and after ten o'clock I neither saw nor heard him.

I continued to watch until 4:08 p.m., or for nearly ten hours. As far as I could tell, only the female incubated. She always sat sideways in her domed chamber, with her head at the left side (from her point of view) of the doorway, her tail turned sharply to her right and pressed against the right edge of the opening, projecting beyond the inner cup but not beyond the outer shell of the nest. Her 11 sessions which I timed ranged from 22 to 39 minutes and averaged 29.9 minutes. Her

11 recesses (after her first session in front of the blind) ranged from 7 to 34 minutes and averaged 18.7 minutes. They became shorter after midday; but even after she had sat for a total of two hours in my presence, she took an outing that lasted 34 minutes, and her return was probably not delayed by continuing distrust of the blind. She incubated with a constancy of 61.5 per cent.

Nestlings

On the afternoon of 9 June, both eggs were pipped. Late on the following morning, the female sat in the nest until I almost touched her, then dropped into the pathway in front of the nest, over which she crept in a crooked course, with depressed head and body and half-spread, beating wings. She resembled a small creeping mammal. Her departure revealed two nestlings, which were pink-skinned with long but sparse gray down. The interior of each nestling's mouth was deep yellow, and the flanges at the corners were very pale yellow. Their eyes were tightly closed.

I saw the male approach with a billful of insects for the nestlings while their mother brooded, but neither he nor she would feed them while I watched unconcealed. Yet the female continued to sit with amazing steadfastness. One morning when I came to photograph the nest, I found her brooding. She remained while I set my camera on a tripod and made eight time exposures with the lens only 3 feet from her. None of the negatives revealed any movement of her head. After the last exposure, she dropped to the ground and "feigned injury" once more.

One nestling vanished a few days after it hatched. When the survivor was a week old, its plumage began to expand. When I visited the nest 11 days after it hatched, it was no longer there. The pulled-up lining suggested that it had been torn from its nest by a predator.

CHESTNUT-CAPPED WARBLER

Basileuterus delattrii

This attractive wood warbler is about four and a half inches long. In both sexes, the top of the head is chestnut, bordered on each side by a white superciliary stripe which is broad on the sides of the forehead and narrows as it passes backward above the eyes and ear coverts. The lores and a triangular spot behind each eye are black. The ear coverts are chestnut like the pileum. The hindneck and sides of the neck are gray, and the remaining upper parts, including the wings and tail, are dull olive-green. The under parts are wholly yellow. The bill is black; the eyes are dark; and the legs and feet are pale flesh-color.

The Chestnut-capped Warbler ranges from Colombia to western Guatemala and occurs casually in the Mexican state of Chiapas. From sea level along the Pacific coast of Central America it extends upward into the highlands, to about 5,000 feet in Guatemala and 6,500 feet in Costa Rica. In this country it resides also on the higher parts of the Caribbean slope, down to about 2,000 feet, but it is absent from the Caribbean lowlands north of Panama. In the Santa Marta region of northern Colombia, this warbler occurs between 800 and 3,000 feet and is more common in the upper part of this altitudinal belt (Todd and Carriker, 1922:440). In the highlands and upper parts of the Pacific slope of Central America, the Chestnut-capped Warbler is abundant in coffee plantations, hedgerows, scrubby pastures, the bushy sides of ravines, and light woods with much undergrowth. Around San Miguel de Desamparados, Costa Rica, 4,500 to 5,000 feet above sea level, in October of 1935, this was one of the most common birds. At low elevations along the Pacific coast north of the Gulf of Nicoya, it inhabits arid woodland where the low, thorny trees lose much of their foliage in the dry season. Farther to the south, in the Térraba Valley of Costa Rica, Chestnut-capped Warblers of a different race (*B. delattrii mesochrysus*) frequent vegetation that is far more luxuriant. Strictly avoiding the lofty rain forest that not long ago covered most of this valley, as likewise the low, impenetrable thickets which promptly take possession of neglected clearings, the warblers live in vegetation of an intermediate type, in which there are tall second-growth trees, scattered or in open stand, with bushes and tangles of vines beneath or beside them.

I have never found Chestnut-capped Warblers flocking, but they live in pairs at all seasons. Years ago, when I settled down to do some writing on a coffee plantation at the southern edge of Costa Rica's Central Plateau, I spent the greater part of each day at a table before a window that looked out into the coffee bushes, which had the usual light shade of low, pruned trees. It was October, and most of the birds had finished nesting months earlier. Several times each day, a pair of Chestnut-capped Warblers came to forage among the glossy foliage of the coffee bushes that almost touched the side of the house. Often I saw both together; at other times only one came within view, but its liquid calls were answered by a similar voice among the dense shrubs.

A single male Wilson's Warbler, wintering here, kept this pair of resident warblers company; and I repeatedly saw him with them. Such mixed trios were not uncommon in the vicinity. The black-capped warblers were, even in their winter home, intolerant of the presence of other individuals of their species; and the pairs of Chestnut-capped Warblers defended their territory against the intrusion of other individuals of their species; but the former seemed to seek, and the latter to tolerate, the companionship of a different species. These

Chestnut-capped Warblers belonged to the nominate race. Later, in the valley of El General, I learned that the race *mesochrysus* also lives in pairs throughout the year.

In its dietary habits, the Chestnut-capped Warbler hardly differs from many other members of the family which glean insects and their larvae amid the foliage of bushes and low trees.

Voice

When first I heard a Chestnut-capped Warbler sing, in late January on the Pacific slope of Guatemala, I thought that one of the abundant wintering Yellow Warblers was carolling out of season. As in this wide-spread species, the Chestnut-capped Warbler's little verse is bright and cheerful but lacking in power and brilliance. Through most of the day, he sings amid the thickets where he forages, but in the gray light of dawn he mounts to a high, exposed perch to perform. One male sang in the top of a medium-sized Cecropia tree that rose above low, tangled growth. In the open, sparsely branched crown he was conspicuous even in the dim light, as he flitted from limb to limb while he tirelessly repeated his earnest little lay, continuing for many minutes. Another male used to sing on May mornings about 40 feet up in the top of a small, leafless tree that grew in a bushy pasture. He remained more constantly in one spot than the first bird, and often sang continuously for ten minutes or more. In the valley of El General, where these birds sang, the Chestnut-capped Warbler resumes singing in early or mid-February and sometimes still performs freely in June, with sporadic song in July and August.

Nest and Eggs

The Chestnut-capped Warblers' nests are so well concealed that, in over 20 years in localities where these birds breed in small numbers, I have found only five, and all were finished, or nearly so, before I noticed them. All of these nests were on our farm at Quizarrá in the valley of El General, about 2,500 feet above sea level, and all were covered structures situated on the ground. Three of the nests were on, or at the top of, moderate slopes; one was halfway up a steep bank about 40 feet high; and one was on nearly level ground. These nests were beneath light secondary vegetation or in rather open areas not far from dense growth; not the vegetation above them so much as the way they were set amid the ground litter, beside or between rocks, or against a fallen log, made them difficult to detect. One nest was built in a shallow depression in the soil, which the warbler may have scratched out for its reception; it was partly buried in the ground litter and screened by a sedge and other low herbs. Another nest was set in a nook between two partly exposed gray rocks, the tops of which were level with its top. A slender root, bridging the space between the rocks, passed over the roof of the nest, which was nearly covered by big fallen leaves. Yet another nest was placed with its back against a decaying log, in a patch of bracken fern, weeds, and vines, amid which there was much dead vegetation lying on or near the ground. The nest on the high escarpment was hidden in a little niche or recess, beneath a small projecting rock, behind low herbage that completely screened its doorway.

I should not have discovered any of these cunningly hidden nests if a parent had not flown from them as I walked by, or if it had not approached them with building material or food for nestlings in my presence. Even when by these means I learned the approximate location of the nest, it usually took many minutes of the most careful searching to find it. Sometimes it was necessary to watch the warbler leave the nest, or take food to it, twice before I could see it. In 1948, I found my first Chestnut-capped Warbler's nest in a thicket beside a banana plantation. In the following breeding season, the warblers spent much time in the same area and I searched for their nest fruitlessly for several weeks. Then, one day, I watched the singing male fill his bill with tiny insects and descend to the ground, directing me to the spot where his nestlings were hidden, only 5 yards from the site of the earlier nest. I had gone over this ground again and again without finding the new nest.

The Chestnut-capped Warbler builds a roofed or oven-shaped nest, as do all other species of *Basileuterus* whose nests are known to me. The nest of the present species is distinguishable at a glance from that of its neighbor in El General, the Buff-rumped Warbler, by its lower roof and lower, narrower, less obvious doorway. One nest was composed of fine herbaceous stems, grass blades, fragments of dead dicotyledonous leaves, leaf skeletons, rootlets, tendrils, the fungal rhizomorphs called vegetable horsehair, and the like. In the bottom was a thick pad of finely shredded bast fibers. This nest was 4½ inches high, 4½ inches from front to back, and 5½ inches from side to side. The inside measurements were 2½ inches high, 4 inches from front to back, and 3 inches from side to side. The doorway was 1¾ inches wide and 1⅜ inches high.

In a nest, which was practically finished when I discovered it on 20 April, the first egg was not laid until 29 April, and two more followed at intervals of one day. The second egg was laid before 6:45 a. m. Each of two nests contained three eggs, two held two nestlings, and one nest sheltered only a single nestling. The eggs of one set were white, marked with fine specks of cinnamon that were crowded all over the thicker end but sparingly scattered over the remaining surface. They were moderately glossy and they measured 17.9 by 14.3, 17.5 by 14.3, and 17.5 by 14.3 mm.

CHESTNUT-CAPPED WARBLER

In five nests in the valley of El General, 2,500 feet above sea level, eggs were laid as follows: April, 1; May, 3; July, 1.

Incubation

Only the female incubates. From 5:29 to 11:40 a. m. on 16 May 1955, and from 12:12 to 6:12 p. m. on the following day, I watched from a blind a nest that held three eggs well advanced in incubation. The whole morning was sunny, but the sky became overclouded after two o'clock in the afternoon and there was a brief, light shower, followed by negligible drizzles. In somewhat over 12 hours of watching, I timed nine full sessions, ranging from 27 to 70 minutes and averaging 44.6 minutes. There were ten recesses ranging from 16 to 35 minutes and averaging 23.3 minutes. The eggs were covered for 65.7 per cent of the female's active period, which began with her first departure at 5:30 a. m. and ended when she settled on the nest for the night at 5:05 p. m., more than an hour before it became too dark to see her. She incubated somewhat more constantly on the sunny morning than on the mostly cloudy afternoon. Her five recesses in the forenoon averaged nearly the same as the five which she took after midday; but her five morning sessions averaged 50.6 minutes, whereas her four afternoon sessions averaged only 37 minutes. Nevertheless, because she retired in the evening almost an hour before sunset but became active in the morning more than half an hour before sunrise, her total time outside the nest was much greater in the forenoon than in the afternoon.

The warbler who incubated never sang in my presence, but uttered only sharp *chip*'s when approaching or leaving the nest, which seemed to indicate that it was always the female. In the nest, she always sat facing the doorway and was perfectly silent. Her mate stayed most of the time beyond sight and hearing, but thrice he accompanied her as she returned to her eggs. On two of these occasions he came quite close to the nest, but he did not once stand in the doorway to examine its contents.

At 10:40 a. m. the female returned alone to her nest; and while I was recording this fact in my notebook she left again, to perch on a low twig and watch a slender brown snake, possibly a foot in length, glide by about a yard from the nest and vanish into the dense vegetation in front of the blind. After its disappearance, she flew away for a few minutes. (This brief interruption was not counted as a recess.) Evidently she had been alarmed by hearing the snake glide through the vegetation, for she had apparently not noticed it before she entered her nest and she could hardly have seen it through the doorway.

Because of the premature loss of the only nest which I found before the eggs were laid, I could not learn the length of the incubation period.

Nestlings

The newly hatched young are typical wood warbler nestlings, with sparse gray down on pink skin and yellow mouth cavities. From 6:30 to 10:30 on the cloudy morning of 22 July 1948, I watched a nest with two nestlings that were two or three days old. The female brooded for four intervals, lasting 70, 42, 44, and 17 minutes. During the second of her sessions on the nest, a hard shower fell. Both parents fed the nestlings, a total of nine or ten times in the 4 hours. Four times the male took food to the nest while the female was sitting. On three or four occasions she brought food when she came to brood, and on two other occasions the parent's sex was not determined. As far as I could see, the nestlings' meals consisted wholly of insects, among which long, green caterpillars were conspicuous. The parents always approached the nest through the thicket beside and behind it, never directly across the banana grove in front. They always advanced with extreme caution, making their way by hopping from twig to twig, with many pauses to look around, and often oscillating backward and forward as they neared their goal. I heard the male sing only one brief verse all morning. Once as the female neared the nest she repeated many times a sharp *churr*. She always sat facing outward, with her chestnut and white head prettily framed in the round doorway.

When the nestlings were eight days old, their feathers were expanding rapidly. A few days later, their upper parts were uniform dark gray, without the head markings of the adults. Their under parts, as far as I could see without removing them from the nest and possibly hastening their departure, were yellowish olive. Both these nestlings left when 12 days old.

Family THRAUPIDAE

FLAME-COLORED TANAGER

Piranga bidentata

This large tanager is about seven inches in length. In the more richly colored southern forms of the species, the male is orange-red or scarlet on the head, neck, and under parts. The upper parts of his body, behind the head, are brownish red, with broad dusky streaks on the back. The dark grayish brown wings bear two conspicuous pinkish bands on the coverts. The three outer feathers on each side of the grayish brown tail are broadly tipped with white on the inner webs. The female is olive-green above, yellow below, and her back is streaked as in the male.

The Flame-colored Tanager inhabits the highlands from Mexico to western Panama. In the former country it occurs chiefly above 2,500 feet (Miller *et al.*, 1957:305). In Guatemala, where it is said to be fairly common in the pine and oak forests from 5,000 to 9,000 feet (Griscom, 1932:381), I failed to become acquainted with it. In Costa Rica, this tanager lives chiefly in the high mountains from 6,000 feet to timber line, although occasionally it wanders down as low as 3,000 feet. During my year at Montaña Azul, I met it very rarely in the neighborhood of my dwelling at 5,500 feet, but it became more abundant as I climbed upward. I saw far more of the Flame-colored Tanagers at La Giralda at the western end of the Barba massif, where they frequented the tall introduced cypresses and the indigenous shade trees in the pastures between 6,500 and 8,000 feet. They were also found in the upper levels of the high forest that filled the neighboring ravines. When I arrived here toward the end of February, they were already paired. Although these tanagers usually foraged high in the trees, where it was difficult to see what they ate, they sometimes descended to lower levels. Late one morning in April, I found a pair near the ground in a pasture. The female ate a berry of *Satyria Warszewiczii* that she pulled from a bush growing on a stump. Once she alighted momentarily on the grass, apparently to catch an insect.

At La Giralda in late February, the male Flame-colored Tanagers were already singing freely, and they continued, with fluctuations in the frequency of their song, until well into June. They sang high in the shade trees that stood isolated in the pastures, sometimes on the topmost spire of a tall cypress, not only at dawn but through much of the forenoon. The song, like that of the Summer Tanager, was melodious despite a touch of harshness or huskiness in the full, deep notes. One Flame-colored Tanager sang *Chewee-very-vire*. The *very* was lowest in pitch, the final *vire* abruptly higher. Sometimes he sang the same notes in a different order, such as *Chewee-very-vire-very*. There was a richness suggestive of midsummer lushness in his somewhat throaty voice.

The Flame-colored Tanager's call is a full-voiced *prr-rt prr-rt*. While a male sang in one treetop, early in the season, a female in a neighboring tree repeated these call notes over and over. A few days later, I saw a male followed by two females, one of whom flew into the forest with him, while the other remained in a tree in the pasture calling *prrt prr-rt*.

Nesting

In mid-April, I found a female building in the top of a middle-sized tree standing in a pasture. The foliage of this tree was so dense that she quite vanished whenever she flew into it with material in her bill, and all my efforts to locate the nest more precisely were defeated. Among the materials taken to the nest were rootlets of epiphytes that the builder pulled from the trunks of neighboring trees. The building female, who worked sporadically, was often followed by her mate, but I detected nothing in his bill. These tanagers wandered widely and were not often to be found within sight of their nest tree.

At the end of May, I at last found a nest of this species that was such a prominent member of the local avifauna. It was about 25 feet up in the top of a *Viburnum* growing in a pasture with scattered trees. Although the crown of this tree was very dense, the nest was built on a shoot rising above the compact mass of foliage, where is was plainly visible, even at a considerable distance, from the slope rising above the tree. But the parents were always so cautious in approaching their nest that, although I had passed this tree many times, I did not suspect its presence until one morning when I noticed the female at the edge of the neighboring forest and followed her movements. Directed by the hint that she gave me, I found her nest before she went to the nest tree. Indeed, her mate, arriving later, fed the nestlings before she did. He was breeding before he had acquired full nuptial plumage; his head and breast were paler than in mature males, and his abdomen was yellow rather than flame-color. Evidently the male Flame-colored Tanager requires a year or more to develop the definitive colors of the adult.

This nest held two nestlings that were already feathered and left a few days later. By 3 June the nest was empty and I collected it. The open cup was made largely of rather stiff rootlets, chiefly if not wholly from epiphytic plants, and a few fine twigs. The lining consisted principally of the fine inflorescence stalks of grasses. Much light passed through the meshes of the loosely constructed walls. For a tanager, this was an unusually slight nest. It measured 4½ by 4 inches in over-all diameter by 2½ inches in height. The inside diameter was 3 by 2½ inches and the depth 1¾ inches. When I climbed into the *Virburnum* tree to cut down the nest, the male flew close around it and showed much concern, but his mate and the fledglings had already gone so far that I could not find them.

COMMON BUSH-TANAGER

Chlorospingus ophthalmicus

The small tanagers of the genus *Tangara* which forage through the crowns of the lofty trees of the tropical American forests present a variety of color and pattern that is the constant delight to those who watch them. These brilliant, active little birds remind one of the arboreal wood warblers of northern woodlands, and as gleaners of insects in the treetops seem to replace them in lowland forests where warblers occur chiefly as migrants. The multicolored species of *Tangara* are perhaps more abundant in the foothills than on the coastal plains; in parts of Costa Rica they are still quite numerous at an elevation of 4,000 feet, and on the outer slopes of the equatorial Andes one sees many of them considerably higher; but as one ascends to cooler heights they rapidly fall away. In southern Central America, the Spangled-cheeked Tanager is the only species of *Tangara* that lives much above 5,000 feet. In the cool highland forests this beautiful genus is replaced by another group of small tanagers of far plainer attire, the olive-green bush-tanagers of the genus *Chlorospingus*.[1]

One of the most widespread and abundant species of *Chlorospingus* is known as the Common Bush-Tanager. It is a stout bird somewhat over five inches in length, and the male and female are indistinguishable in plumage. The top and sides of the head and neck are (in the Costa Rican race *regionalis* with which this account is chiefly concerned) sepia-brown, grayish brown, or sooty brown, with a prominent white spot behind each eye. The remaining upper parts are plain olive-green. The chin and throat are white tinged with brown or buff; the chest, sides, flanks, and under tail coverts bright olive-yellow; the breast and abdomen white. The short and rather thick bill is blackish; the eyes brown; the legs and feet dark.

This widespread bird ranges through the mountains from southern Mexico to Bolivia and northern Argentina, and its altitudinal distribution varies regionally. In Mexico it is known to occur, in its vari-

1 "Bush-tanager" is not an appropriate name for these tanagers, which often forage high in trees. I propose the name "verdiger" for the consideration of future revisers of vernacular nomenclature. The present species would become "Common Verdiger."

ous races, from 11,500 to 2,500 feet above sea level, and in the winter one form descends into the lowlands of the State of Veracruz (Miller *et al.*, 1957:310-311). In Guatemala, I have found it up to nearly 10,000 feet, and am not certain whether it extends higher. In Costa Rica, it has been recorded by Carriker (1910:840) as low as 1,200 feet in the cool, humid forests of the Caribbean slope; but on the drier and sunnier Pacific slope I have not seen it below 3,500 feet, in the southern part of the country. On both sides of the Cordillera, it is abundant between 4,000 and 8,000 feet and found occasionally about 1,000 feet higher. From 6,000 or 7,000 feet upward, the Common Bush-Tanager occurs in the same localities as the Sooty-capped Bush-Tanager. In Venezuela, the several races of the Common Bush-Tanager have been found from 3,000 to 10,000 feet above sea level (Phelps and Phelps, Jr., 1963: 388-389).

The Common Bush-Tanager dwells in the humid subtropical forests, where every larger tree is heavily laden with epiphytes that range in size from trees and large shrubs through orchids, bromeliads, and ferns of bewildering variety down to the mosses and liverworts which envelop trunks and boughs with thick verdant cloaks and lichens which cling to patches of bark that the bryophytes have failed to cover. These forests often have a dense undergrowth of cane-like bamboos or broadleafed shrubs which, on precipitous slopes, distressingly trammel the movements and obstruct the vision of the ornithologist who tries to keep sight of small, obscurely colored birds that flit through the encumbered boughs far above him. Only when they forage at the forest's edge, or venture forth into the scattered trees and bushes of an adjoining clearing, do bush-tanagers permit the watcher to see much of themselves. Although by no means restricted to the interior of the heavy montane forests, these greenish tanagers seem not to thrive far from it; they follow its retreat from the centers of human population, and even at suitable altitudes are rarely seen in districts that have been long deforested.

In September or October, after the young of the year have become self-supporting, bush-tanagers unite in small flocks, sometimes containing a dozen or so, that wander through the woodlands in company with other small birds. There is no indication that pair bonds are preserved within the flocks. In the Guatemalan highlands, I so often saw Common Bush-Tanagers with Golden-browed Warblers that I concluded that the bright warbler and the dull colored tanager had a special attraction for each other. I found them together on the Sierra de Tecpam above 8,000 feet, and also on the Volcán Zunil between 5,000 and 7,000 feet, where both species were very abundant. At around 7,000 feet in the Costa Rican mountains, I found in early March a pair of bush-tanagers in a flock that also contained two Ruddy Treerunners, two Yellow-thighed Finches, a wintering Golden-winged Warbler and a wintering Black-throated Green Warbler. The

FIG. 6. Subtropical forest on the northern slope of Costa Rica's Cordillera Central, near Vara Blanca. The pasture in the foreground, with stumps densely covered by epiphytes, is at about 6,000 feet. In this forest and at its edges lived Common Bush-Tanagers, Yellow-thighed Finches, Chestnut-capped Brush-Finches, and Yellow-throated Brush-Finches. July, 1938.

next day I saw a single bush-tanager with a Collared Redstart, a Black-cheeked Warbler, a Flame-throated Warbler, and a Spotted Barbtail. Probably each of these birds had a mate that I failed to detect amid the heavy vegetation. Near the lower limit of its vertical range, the bush-tanager sometimes associates with its brighter relatives, such as the Silver-throated Tanager, the Bay-headed Tanager, the Golden-masked Tanager, and the White-winged Tanager.

FOOD

Like other small tanagers, bush-tanagers subsist on a diet containing large portions of both animal and vegetable foods, and on occasion they drink nectar. They forage among the foliage and the moss-covered branches in much the same fashion as an arboreal wood warbler, sometimes on the lower boughs, sometimes high in the treetops. Often they hang upside down to reach an insect or spider. Like many other birds, they are most likely to forsake the dark forests for open places during long-continued wet spells. At Montaña Azul on the storm-beaten northern slope of Costa Rica's Cordillera Central, a small flock of these warblers would sometimes come on dismal, drizzly days to forage around the cottage, where there were no trees. In such weather a party of ten or more might appear repeatedly in the course of a day, announcing their arrival by the low, weak chirping that they kept up continuously as they hopped around searching for insects. They hunted through the dense, low hedges of trimmed cypress trees that surrounded the house, and they even dropped to the ground, to hop over the lawn or the bare soil of flower beds, in company with such dooryard birds as Yellow-faced Grassquits and Rufous-collared Sparrows. Then, having made the round of the hedges, they flew off into the chill gray cloud-mist that enveloped us, chirping lispingly as they vanished.

Among the fruits which the bush-tanagers eat are the berries of *Conostegia* and other arboreal melastomes. They peck into the juicy fruits of *Hedyosmum Artocarpus,* which are far too large to be swallowed whole. The berries of *Satyria Warszewiczii,* a vigorous shrub of the heath family that forms great compact clusters on trunks and stumps in Costa Rican mountain pastures around 7,000 feet above sea level, are especially attractive to these tanagers. This shrub has an interesting adaption which makes it more convenient for the various kinds of birds that pollinate its flowers or disseminate its seeds. The long, slender, tubular flowers, red with a white tip, are displayed, in dense compact racemes, inward from the foliage on slender horizontal or descending branches. These flowers generally point downward, with the opening toward the ground—an arrangement which seems necessary to permit the pollen to escape through the pores at the tips of the anthers—so that the Purple-throated Mountain-gem, the Green Violet-ear, and other hummingbirds which sip their nectar must hover below them with nearly vertical bodies in order to push their long bills upward into the floral tubes. The *Satyria* blossoms chiefly in the drier months of February and March; and the fruits, about half an inch in diameter, ripen in quantities in May and June, when they provide food for nesting birds. The maturing berries turn from green to white, then pale violet, deep violet, purple, and finally almost black, and as their color deepens they become juicy and develop a pleasant sweetish-acidulous flavor, attractive to small boys (who call them *muelas*) as well as to a variety of birds. As the fruits ripen, the clustered short pedicels that bear them, and were downwardly directed at the flowering stage, curve upward, elevating the fruits to the level of the branch, where they are readily accessible to perching birds that cannot, like the hummingbirds, hover motionless while gathering their food. Other birds which eagerly seek these juicy fruits are Golden-browed Chlorophonias, Flame-colored Tanagers, Yellow-thighed Finches, and Prong-billed Barbets. All of these birds, like the bush-tanagers, take a whole berry in their mouth, then by skillful manipulation with their bill remove the contents from the rather tough skin, which they drop to the ground. The juicy pulp is then swallowed with its minute embedded seeds, which the birds spread widely over the mountains.

When gathering fruits, no less than when searching for insects, bush-tanagers occasionally descend to the ground, as I have seen them do to pluck the little orange berries of *Gomozia granadensis* (also widely known as *Nertera depressa*), a slender creeping herb abundant in the high mountains of tropical America.

On the Sierra de Tecpam in 1933, I saw little of the Common Bush-Tanager during the dry months early in the year; but in July they became numerous, especially in low, broadleafed woods that had grown up in openings in the cypress forest around 9,500 feet, and they remained abundant until December. When, in the sunnier weather toward the end of the year, the shrubby *Salvia nervata* began to display its crimson blossoms profusely in the more open parts of the woods, the bush-tanagers often visited them for nectar. Since these short-billed birds could neither probe the long, slender corolla tubes as the hummingbirds did, nor puncture their bases in the manner of their neighbors the flower-piercers (*Diglossa*), they resorted to yet another method of procuring the sweet fluid. Perching near an inflorescence, they pulled a corolla from its calyx, mandibulated the basal end, doubtless to press out the nectar, then dropped the corolla. Sometimes after discarding it they probed with bill or tongue into what was left of the flower.

After the birds finished their feast and flew away, I examined the ground beneath the *Salvia* shrubs and found it strewn with hundreds of blossoms. The majority of the fallen corollas had been torn off just above the calyx, leaving a stump attached to the plant, but many had been pulled away entire; that these last had not fallen spontaneously was obvious from the marks which the birds' bills had impressed upon the delicate tissues. Probably when the bush-tanagers left the stump of the corolla tube in the calyx, they probed it to obtain the nectar that remained there; but when they pulled away the corolla entire, they pressed the nectar from its base. They did not permit me to come close enough to make sure that this was the reason for their two modes of procedure. Slud (1964:366) reported seeing the Common Bush-Tanager "chewing" flowers in Costa Rica.

VOICE

Bush-tanagers do not, like thrushes, compensate for dull plumage with brilliant song, and vocally they are no more gifted than the majority of their colorful cousins of the genus *Tangara*. As they fly about in pairs at the approach of the breeding season, or pursue rivals, they rapidly repeat sharp, twittering notes. This tuneless utterance should doubtless be considered their song, for at dawn in the nesting season it is repeated over and over, with that persistence whereby poor songsters, such as the American flycatchers, seem to try to compensate with quantity for what they lack in quality. In a shady pasture high on a mountainside at the end of March, I heard a bush-tanager perform in this uninspiring fashion for half an hour. In the first dim light of dawn, he began to sing while hidden in a tree with dense foliage. As it grew lighter, he continued with hardly a pause while he flew from tree to tree in his territory, sometimes with his mate. His rapid, sharp twitter was repeated over and over at short but variable intervals. There was also some variety in the form of his utterance. Sometimes the dry notes came more swiftly toward the end of a series. At times his ticking resembled the song of the Golden-masked Tanager, although the notes were higher and sharper. At other times the bush-tanager's utterance might have been mistaken for the twittering of an excited hummingbird.

Other notes that I have heard from Common Bush-Tanagers are weak *chirp*'s and *chip*'s—companionable notes which they voice as they flock together after the close of the breeding season, especially in inclement weather. Moynihan (1962) has analyzed in great detail the apparently simple and limited vocabulary of the bush-tanagers, and distinguished a variety of utterances appropriate to the different occasions in their lives.

Territory

As the nesting season approaches, the small winter flocks of the bush-tanagers split into pairs, a change in their social relations which occurs in March, if not earlier. Sometimes the establishment of breeding pairs is attended by difficulties, for two individuals aspire to the same mate, or the same territory. In a shady pasture at about 7,000 feet at La Giralda, I found on 12 March three bush-tanagers who pursued each other from tree to tree, rapidly repeating the usual sharp twitters as they flew. These pursuits continued for about half an hour, with intermissions in which all three ate berries close together. Once two of the birds clutched each other in a tree and fell to the ground, where they promptly separated. Unfortunately, I could not distinguish the sexes of the contestants. For the next four or five days, the trio behaved in much the same manner, sometimes continuing their chasing for three quarters of an hour; but I saw no more grappling encounters. I have never known bush-tanagers to engage in such fierce conflicts as Moynihan (1962:318) has described, in which two birds cling together, pecking and biting furiously, striking each other with feet and wings, until the contestants fall to the ground, tumbling over and over, still fighting as they fall.

By 19 March only two bush-tanagers remained in possession of the shady pasture. They flew from tree to tree over an area about 100 yards long by half as wide, and they were much more silent than they had been while the third bird was present. The male sang in these trees at dawn, as earlier recounted. But I could find no nest here until two months later, on 21 May, when I discovered a female building in a tree at the edge of the area over which the pursuits had taken place.

Possibly this was a second nesting of this pair, and an earlier one had escaped my notice.

Nest Building

Few birds choose a greater variety of sites for their nests than the Common Bush-Tanagers, which may build in the midst of the forest or in a bushy pasture, and at all levels from the ground to at least 50 feet up in a tree. Of the nine nests that I have seen, six were in trees, hidden among the moss and larger epiphytes that grow profusely on trunks and limbs in the humid mountain forests. At times the structure is embedded in the thick layer of green moss that covers an upright trunk. All that can be seen from the ground is a small round opening amid the moss; one would never suspect the presence of the nest without hearing the cries of the hungry nestlings within, and even then he is not likely to discover just where they are until he has seen a parent enter with food in its bill. Often the builders seek additional protection from a larger epiphytic plant. A nest in the high mountains was 40 feet up on a horizontal limb of a tall tree in a pasture, beneath a big tank bromeliad whose broad leaves formed a roof above it. Another nest was attached to the base of a large *Tillandsia* with long pointed leaves, growing at a height of 10 feet on the upper side of an ascending mossy bough of an old, spreading *Viburnum* tree standing in a pasture, not far from forest. Still another nest was 40 feet up in a large clump of hart's-tongue fern (*Elaphoglossum* sp.) perched on a slender, ascending bough of a middle-sized tree in the midst of the forest. This nest was placed in a cavity among the rhizomes of the fern, the moss which had overgrown them, and the dead leaves that had lodged among them. The builder completely vanished each time she entered the cavity. This nest was near Cañas Gordas in extreme southern Costa Rica, at an altitude of only 3,500 feet, the lowest point at which I have found the bush-tanager breeding—or attempting to breed, as the nest was apparently never used.

In the same locality where some bush-tanagers built beneath epiphytic bromeliads, another pair set two successive nests in deep niches in the creeping grass that densely covered a vertical bank beside a cowpath in an open pasture, about 75 feet from a small stand of second-growth woods. The bank was 4 feet high, and the nests were situated near its top. Lower on the same mountainside, I found another nest on gently sloping ground, at the base of a small shrub amid dense, low grass that concealed it well. This site in a bushy pasture was about 150 feet from a patch of woods. Although seven of the nine nests that I have seen were in highland pastures, all were a short flight from woodland where the tanagers foraged. In the Province of Chiriquí in western Panama, Worth (1939) found a nest 9 feet up in a tall coffee bush in a plantation.

The only nest of the Sooty-capped Bush-Tanager that I have seen was at the top of a roadside cut-bank 15 feet high, where it was well concealed by ferns and small-leafed heaths growing at the edge of the bank. Aside from the Common Bush-Tanager, the only member of the family that I have known to nest on uniformly sloping ground is the Blue Tanager, which does so with great rarity, most of its nests being in shrubs or trees, sometimes high. The range of sites chosen by the Common Bush-Tanager is almost equalled by those of its neighbor the Flame-throated Warbler, which in a single locality may shelter its nest beneath tank bromeliads in a treetop or hide it in a grassy bank in a pasture.

As far as I have seen at three nests, only the female bush-tanager builds. The earliest nest that I have known was that at 3,500 feet above sea level in southern Costa Rica. When found on 17 March 1964, the supposed female was actively building, collecting moss, bits of fern fronds, fibers, and the like from branches of trees well above the ground and taking them into the deep recess in the clump of hart's-tongue fern. In a half-hour in the middle of the morning, she took 17 billfuls to the nest site. After this, I looked in vain for the tanager in the vicinity of her nest. Possibly her undertaking was abandoned because the dry season that year was severe and prolonged, continuing well into April.

On 12 May 1963, I found a bush-tanager building the above-mentioned nest beneath a *Tillandsia* in a *Viburnum* tree. She was not shy and continued to work while I sat in the pasture at no great distance. From 8:48 to 9:48 a.m. she brought material 26 times, and in the following half-hour she did so 11 times. Usually she came with a great billful of green moss, which she had plucked from the nest tree itself, although sometimes she brought moss from a more distant tree. While clinging to an upright trunk to pull off moss, she frequently flipped her wings, flashing the dull white of her under wing coverts, which caught the observer's eye when the olive-green bird herself was difficult to detect against the dark green background, in the shade of the tree. Sometimes the builder dropped to the ground to gather dry fragments of grass for her nest. Often she stayed in the nest for from half a minute to a minute, while she arranged her materials and shaped it.

The following morning, this tanager was slow in starting her work. From 6:25 to 7:25, she made only six trips to her nest. From 7:33 to 8:33, she brought 24 billfuls of materials. Then she reduced her pace and came with only 13 billfuls in the next hour. Now she brought no green moss, but only dry materials for the nest's lining. Sometimes she collected bits of dry grass blades from the pasture, often plucking them while she clung to the base of an upright stem of weed or shrub, although at times she stood in the grass. She also gathered fibers and other fragments from the decaying stems of epiphytes, well above the ground. Whatever material she collected, she usually brought a sheaf of it, rather than a single piece. When she arrived at the nest, she might deposit her load and promptly leave, or she might remain for from 10 to 30 seconds, and rarely for over a minute, while she shaped her structure. Once she flew at a Spotted-crowned Woodcreeper who was hunting along a limb of a neighboring tree. While working, she at long intervals uttered the usual twittering refrain, and at times she voiced a little *chip*, like that of the wintering Wilson's Warblers that were so abundant in the vicinity, but weaker. When I stood beneath her nest, she complained with a rapidly repeated, sharp, clicking note.

While this female bush-tanager built, I saw and heard little of her mate. Once he clung to the branch beside the nest, holding a rootlet in his bill and flipping his wings. While he was there, his mate entered. Finally he flew away, without taking the rootlet into the nest. On another occasion, he pursued the female as she carried a contribution to the nest.

The third nest that I watched during construction was not found until the female had begun to line it with fragments of dead vegetation gathered from epiphytes growing on the lower part of the nest tree itself. Chiefly she tore flat, brown fragments from the bases of dead outer leaves of bromeliads. She filled her bill with a number of these pieces before ascending to the nest. The following morning she brought finer materials, including rootlets or the like, which with great effort she tore from the epiphytic vegetation on mossy trunks. On the first morning, she made nine trips to the nest in 25 minutes; on the second, 12 trips in an hour. Her mate accompanied her to the nest a few times, but mostly she came and went alone. Once he seemed to pick a fragment of material from a branch and take it to the nest, but he did not enter with it. In contrast to the two females who built in trees, paying little attention to their watcher, the tanager who built on a grassy bank would not approach her nest while I sat, partly concealed, about 75 feet away.

Completed nests were bulky open cups, depending for protection from the cold mountain rains on the dense vegetation amid or beneath which they were set. One nest, however, was thinly roofed with loose, detached pieces of green moss, which seemed to have been brought for this purpose by the builder, although it was possible that they had fallen from the branches above, which were heavily swathed in the same kind of moss. This nest, which had a wide doorway on the outer side, was composed principally of moss, and the bottom was well lined with fibrous materials. Other nests were of more varied composition. That beneath the *Tillandsia* had an outer layer of green mosses and liverworts. Next within was a layer in which grass blades predominated, intermixed with black fibrous rootlets and moss setae with the capsules still attached. The lining was of horses' or cows' hairs, fine rootlets, many moss setae, and fine flat fragments of vegetation.

Bulkiest of all were the nests set in creeping grass on a bank. The massive foundation consisted of a great pile of coarse materials, including grass blades, herbaceous stems, long and rather thick rootlets, fragments of fern fronds, whole small dicotyledonous leaves and pieces of larger ones, all loosely piled together. The outer layer of the cup itself consisted of similar materials more firmly compacted and mixed with many green mosses and liverworts. The middle layer of the cup was composed largely of grass blades or bamboo leaves, becoming progressively smaller inward, and interspersed with fine rootlets. The inner lining was of many fine rootlets and fungal rhizomorphs, mostly blackish, and other fibrous materials. Many of the grass blades and dicotyledonous leaves had been added to the nest while still fresh and green. These bulky nests were 5 to 6 inches in diameter by 4 to 5 inches high. The cavities were 2 to 2½ inches in diameter by 2 to 2¼ inches deep.

The only other species of *Chlorospingus* whose nest I have seen is the Sooty-capped Bush-Tanager. This nest, situated at the top of a 15-foot-high cut-bank as mentioned above, was also a roomy open cup. It was composed chiefly of green leafy liverworts, lined with fine black rootlets, and within this an innermost layer of fine, light colored vegetable fibers. This nest was discovered at Vara Blanca, Costa Rica, at an altitude of 6,000 feet, on 7 July 1940, when it contained two nestlings that were already feathered and resembled their parents.

The Eggs

The Common Bush-Tanager appears regularly to lay two eggs: from Costa Rica I have records of two nests with two eggs and of three nests with two nestlings; and this was the number of eggs in a nest found in Panama by Worth (1939). At two nests, the first egg was laid about three days after the completion of the structure, and the second was laid on the following day. The eggs are bluntly ovate, dull white, marked with shades of cinnamon and brown in various patterns. In one set, one of the eggs was closely and finely speckled all over with pale cinnamon, with the flecks somewhat more concentrated on the thick than on the thin end. The companion of this egg bore darker, heavier marks of cinnamon-brown on the thick end, but only a few scattered spots of the same color on the thinner end. In another set, both eggs were spotted and blotched with rusty brown and chocolate, chiefly the former, heavily on the thick end but sparingly over the remaining surface. The eggs of the first set measured 20.1 by 16.0 and 20.5 by 16.0 mm. Those of the second set measured 21.0 by 16.0 and 20.9 by 15.8 mm.

In the Los Cartagos-Vara Blanca region of the Cordillera Central of Costa Rica, 5,500 to 7,000 feet above sea level, eggs were laid in eight nests as follows: April, 3; May, 3; June, 2.

Incubation

After incubation had been in progress for four days in the nest beneath the *Tillandsia*, I spent a morning watching it. A cold northeast wind blew briskly all morning, often driving the cloud-spray through the nest tree, where the nest itself was sheltered from the gale by the bromeliad behind and above it. From time to time a little sunshine filtered through the clouds. Only the female incubated. Between sunrise and noon she took seven sessions on the eggs, ranging from 19 to 39 minutes in length and averaging 30.6 minutes. An equal number of recesses ranged from 14 to 32 minutes and averaged 19.9 minutes. The tanager covered her eggs for 61 per cent of the morning. From my post up the grassy slope, almost level with the nest, I could see only her head, of which the most prominent feature was the white spot behind each eye, shining out distinctly in the deeply shaded cranny beneath the *Tillandsia* plant. On leaving her nest, the female usually dropped downward, not so sharply as some arboreal birds whose nests are higher, and flew for some distance low over the pasture grass before rising into a tree. On her return she was only once followed by her mate, who escorted her almost to the nest. On two or three other occasions, I heard his voice in neighboring trees as the female went back to her eggs. Otherwise, he remained beyond my ken.

At this nest the incubation period was around 14 days. At a neighboring nest it was a few hours more than 14 full days.

Nestlings

Recently hatched nestlings have pink skin shaded by sparse, but rather long, dark gray down. Their eyes are tightly closed. Their mouth has a red lining and is bordered by conspicuous yellow flanges. When the nestlings are about five days old, their eyes begin to open and their pinfeathers to sprout through the skin. At this stage their legs and toes are pink, with pale yellow toenails. At about seven days, the contour feathers and remiges begin to escape from the ends of their long sheaths, but the tail feathers are still completely enclosed. At 10 or 11 days the nestlings are fairly well clothed with plumage and they have expanded remiges, but their tails are still rudimentary. Their originally pink legs and toes have now turned gray.

When five days old, the two nestlings beneath the *Tillandsia* were fed 29 times between 5:45 and 10:00 a.m. Thrice the male brought food while his mate was brooding, and four times the two parents arrived with food almost simultaneously; so that I was sure that both participated actively in this work, although usually I could not distinguish them. Unlike their neighbors in the highland forests, the Golden-browed Chlorophonias, the bush-tanagers did not regurgitate their nestlings' food but carried it in their bills. Often the articles which projected from the parents' bills were green and seemed to be caterpillars, but they were too mangled for positive identification. The nestlings' droppings were usually carried away in the parents' bills; but once the female swallowed a dropping when, after feeding, she stayed to brood.

Although the morning was cool, with little sunshine, the five-day-old nestlings were brooded for only 72 minutes during the 4¼ hours that I watched. The nine sessions of brooding ranged from 1 to 17 minutes and averaged 8 minutes; but the eight intervals of exposure were much longer, lasting from 5 to 52 minutes and averaging 21.8 minutes. If the male arrived with food while the female was brooding, she left. He did not stay to cover the nestlings after delivering what he had brought. I never saw one partner replace the other in the nest, whence I concluded that only the female brooded, as only she had incubated.

A few days later, these nestlings were torn from their nest beneath the *Tillandsia* by some predator. When I visited the nest on the ground beneath a bush at noon on 27 June, one and then the other nestling rushed out while I was still several feet away. Although only 11 or 12 days old, they were fully feathered; they fluttered rapidly over the grass and could even fly a few feet. One jumped up to a twig a few inches above the ground and perched there. It twittered sharply and rapidly, much as adults do, but in a weaker voice. When the nestlings fled from their nest, the parents came very close to me, but they gave no distraction display. Such displays rarely occur in tanagers, and the few that I have seen in other species were rudimentary. These parent bush-tanagers alighted on or very near the ground to feed their fledglings.

In the second nest on the grassy bank, the two nestlings remained until 13 days old. Fledglings closely resemble their parents, with similar white spots behind their eyes.

Second Brood

In the Costa Rican mountains, the Common Bush-Tanagers may rear two broods. In the first nest on the grassy bank, two young were fledged between 23 and 26 May. On 31 May I found the parents beginning their second nest in a similar site, about 20 feet from the first nest. This nest was completed by 4 June, and eggs were laid in it on 7 and 8 June. During the period of building, I saw near the nest a pair of bush-tanagers with well-grown young, doubtless those reared in the first nest. Two fledglings left the second nest on 5 July. The nest on the ground was also evidently for a second brood, for near it, on the day the nestlings hatched, I watched a full-grown young bird beg before an adult who held a berry in his bill.

Family FRINGILLIDAE

YELLOW-BELLIED SISKIN

Spinus xanthogaster

The male Yellow-bellied Siskin is largely black, with yellow on the basal half of all but the central tail feathers; yellow on the breast, abdomen, and under tail coverts; and on the wings a yellow band across the bases of most of the remiges, showing as a small patch when the wing is closed but conspicuous in flight. The sides and flanks are clouded with olive and dusky. The female is olive-green above, with her blackish wings and tail marked with areas of yellow similar to those on the male, but smaller. Her under parts are dull yellowish tinged with olive, becoming more grayish on the throat and fading to dull whitish on the abdomen and under tail coverts. Both sexes are somewhat under four inches in length and have dark eyes, black bills, and blackish legs and toes.

The Yellow-bellied Siskin ranges through the highlands from Costa Rica to Bolivia and western Venezuela. In the latter country it has been found from 2,625 to 6,550 feet above sea level (Phelps and Phelps, Jr., 1963:412). In Costa Rica it occurs chiefly between 6,000 and 8,000 or 9,000 feet. Although Carriker (1910:914) recorded its presence at 2,000 feet, it is certainly rare at so low an altitude. The lowest point at which I have seen the Yellow-bellied Siskin was at about 3,800 feet in the Cañas Gordas district. When I visited this region briefly in mid-January of 1964, a large flock rested in the crowns of tall trees in the pasture at Loma Linda, singing and chattering through much of the day. This social singing was of the usual siskin type, a rapid sequence of small notes poured forth in no particular order and following no musical theme—a chattering kind of song, agreeable but far from brilliant. Sometimes a male chased a female, and sometimes many of the siskins took wing in a compact flock. When I returned to this locality on 14 March of the same year, the siskins had completely vanished; I saw none until 15 June, when a lone male appeared in the pasture where so many had been in mid-January. At an altitude of 5,500 feet at Montaña Azul, I did not see this bird until nine months after I settled here, in April, when a pair arrived to breed. After the failure of their second nest a month later, they left us.

NESTING

The caretaker at Montaña Azul had set a number of long, slender, freshly cut branches of the poró (*Erythrina* sp.) in the little coffee plantation behind his cottage, so that they might take root and sprout and grow into trees to shade his poor, little, moribund coffee bushes, bravely trying to grow in a region too high, cool, and wet for them. Toward the end of April, a pair of siskins arrived and started a nest in the midst of the many shoots that had sprouted from the top of one of these living posts, at a height of 9 feet above the ground.

On the morning of 25 April, beneath the fine drizzle that descended from the low-drifting clouds, I watched the female siskin build amid the fresh young foliage of the clustered sprouts. She was bringing fine rootlets and other fibrous materials to line her little cup-shaped nest. Between 8:30 and 10:00 a.m. she came with 22 billfuls. Although the black and yellow male brought nothing, he was most attentive to his olive-green mate, followed her closely on her journeys to and from the nest, and rested close by, usually preening his feathers, while she shaped her structure. Indeed, his presence seemed indispensable for the progress of the undertaking. Once, when he delayd to escort the female back to the nest, she waited near it with her billful of material until he arrived, and only then went to place it in the nest. On the only other occasion when the male failed to accompany his partner back to the nest, she flew off in search of him, still bearing the rootlet that she had brought for her structure but would not deposit there in his absence.

Once the female flew down to the bare ground between the little coffee bushes to collect material. While she gathered a good billful, the male tugged at the fine rootlets of an uprooted weed. But when the female, having completed her load, flew up to the nest, he followed with empty bill. Although he was not wholly devoid of the impulse to build, this impulse was too weak to be effective.

Beginning on 29 April, the female laid an egg daily until she had three. After laying the second, she started to incubate. On 2 May the eggs mysteriously vanished. On 19 May we discovered that the female siskin was finishing a new nest, 8 feet up in a small cypress tree, 100 feet from her ill-fated first nest. On 21 May she laid an egg and started to incubate, and on the following day she laid her second egg, which seemed to complete the set. Two days later the nest was empty, the eggs having vanished as inexplicably as those in the first nest.

The second nest was a shallow, thick-walled cup, irregularly covered on the outside with tufts of green moss. The substantial wall was composed of fibrous rootlets, the black fungal strands often called "vegetable horsehair," shreds of fibrous bark, and pieces of gray beard-lichen (probably *Usnea* sp.). In the inner half of the thick wall, fibrous bark predominated. The cup was thickly lined with a great quantity of fine, black, fungal filaments. The nest measured 3½ inches in diameter by 2½ inches in height. The diameter of the cavity was 1¾ inches and its depth 1⅛ inches.

The eggs, which vanished before I could measure them, were white with a slight greenish tinge, unmarked.

LESSER GOLDFINCH

Spinus psaltria

The Lesser or Dark-backed Goldfinch varies considerably in coloration in different parts of its wide range. In the southern race, *S. psaltria colombianus,* the males are glossy black on all their upper parts, including the wings, tail, and sides of the head and neck. There are conspicuous areas of white on the wings; white may be either present or lacking on the inner webs of the outer tail feathers. The under parts are everywhere bright yellow. The female is olive-greenish above the olive-yellow below, with white areas on her wings much as in the male. This little finch is only about four inches in length.

The Lesser Goldfinch ranges from Oregon, Colorado, and Texas to Peru and Venezuela. In western Guatemala, I found it from about 3,000 feet near Colomba in the Department of Quezaltenango to the plateau near Tecpam at 7,000 feet. It was abundant in the weedy and bushy fields beside Lake Atitlán in October of 1933. In this country, others have recorded its presence from low elevations on both coasts up to middle altitudes. In Costa Rica, the Lesser Goldfinch is rarer and has a more restricted range. Its recorded occurrence is from somewhat under 3,000 to 6,000 or 7,000 feet above sea level. In all my years in this country, I have seen it in only three localities, all between about 2,800 and 4,000 feet: near Cañas Gordas in the extreme south; on the northern side of the valley of El General at the base of the Talamancan Cordillera; and above Turrialba along the sides of the gorge of the Río Reventazón. Although the Lesser Goldfinch has been repeatedly recorded on Costa Rica's Central Plateau, in a total of a good many months spent at various points in this region I have failed to find it. Evidently these goldfinches, like related species, wander widely when not attached to their nests; but their movements in tropical America have not been studied. Although I was in the Cañas Gordas region continuously from mid-March onward, I did not notice the Lesser Goldfinch until 21 May, when I saw a pair. In June the species became more abundant.

Lesser Goldfinches frequent open country with plantations, scattered trees, and bushy growth. They descend into weedy fields to forage near the ground. Not highly gregarious, they usually travel in pairs or small flocks. They fly with the vertical undulations typical of their tribe, but their dips seem shallower than those of the American Goldfinch. They fed upon the orange flowers of the silk-bark oak (*Grevillea robusta*) planted to shade the coffee plantations in Guatemala (Griscom, 1932:357). I have seen them eat the seeds of a beggar's-tick (*Bidens* sp.) in a weedy dooryard, but have no other observations on their diet. They are wary birds, difficult to watch closely.

Voice

In Costa Rica, Lesser Goldfinches sing and nest at seasons when scarcely any passerines, and few birds of other orders, engage in these activities. In the valley of the Río Buena Vista on the northern side of the basin of El General in 1936, I first became aware of the goldfinches' song in early October, more than ten months after I settled there, and I heard them until early in the following February. Then, just as the majority of the birds were beginning to sing more freely, the goldfinches fell silent. Possibly they left the neighborhood; I have no further record of them until the following June, when I heard them singing again, at a time when many other birds were finishing their breeding and becoming silent.

The full song of the Lesser Goldfinch is a splendid effort, a long-continued, animated outpouring of clear and varied notes that rise and fall in pitch in a most pleasing fashion. It reminds me of the song of the American Goldfinch, and at times, with its wide range of notes, of the most inspired outpourings of the White-collared Seedeater. Sometimes the Lesser Goldfinch's song includes many notes with the same "tearful" quality as its call, which is one of the most sweetly melancholy of bird notes that I know. When at his best, however, the goldfinch uses few of these sad notes that seem incongruous with the blithesome mood of his performance, which is improved by this omission.

Nesting

The late nesting typical of goldfinches in the north is exaggerated in Costa Rica, where the climate permits breeding even by some very small birds at all times of the year, and fluctuations in the abundance of food appear to be chiefly responsible for restricting the reproductive effort of the majority of species to the season most favorable for nourishing the young. At Rivas in the valley of the Río Buena Vista around 2,900 feet, I found three nests: one with eggs in August of 1936; another with eggs in December of the same year; and one under construction on 10 January 1938. On the steep slopes above the Río Reventazón on the opposite side of the country, at an altitude of 2,800 feet, I discovered a nest in which incubation was in progress on 13 September 1941.

The three nests at Rivas were all in roadside or trailside trees of guachipelín (*Diphysa robinioides*), a leguminous tree often planted for living fence posts, whose foliage, as its specific name implies, resembles that of the yellow locust (*Robinia Pseudacacia*) of eastern United States. The three nest trees grew in the valley bottom, amid small coffee plantations, weedy fields, and dense thickets. The nests in these trees were at heights of 12, 15, and 20 feet above the ground. The fourth nest, above the Reventazón, was about 11 feet up in a tuete (*Vernonia patens*), a weedy shrub of the composite family that grew in a scrubby hillside pasture. On the crest of the Western Andes of Colombia, at 6,500 feet above sea level, Miller (1963:58-59) found four nests of this goldfinch, all in isolated cypress trees in a garden, at 3, 4, 6, and 8 feet above the ground.

An unoccupied nest that I pulled apart was a compact, thick-walled open cup. The outer layer was composed of bits of grasses and decaying weed stems, among which were a white feather and some fragments of moss. The middle layer consisted of coarsely fibrous materials, including rootlets and fibers from disintegrating herbaceous stems. The thick lining was of horsehairs, fine vegetable fibers, a few rootlets, and brown, thread-like rachises of the leaves of *Acacia angustissima*. This nest measured 3 inches in diameter by 2 inches in height; the interior was 1¾ inches in diameter by 1⅛ inches deep.

At the nest found under construction in January, I watched briefly and saw the female bring material six times. Although her mate accompanied her on her flights to and from the nest, he brought nothing. While she arranged her materials, he sometimes perched in the nest tree itself and sometimes in neighboring trees, where he uttered only mournful snatches of song, never singing profusely, as he did in the early afternoon.

The eggs in these nests were white, unmarked. The first nest contained three, two of which hatched on 27 August; the second held four eggs on 30 December; and the nest above the Reventazón had four on 13 September. My attention was directed to the first nest by the low, sweetly appealing calls which the incubating female uttered while her mate flitted nervously between neighboring trees, hesitant to approach her in my presence. Soon, however, the pleas of his little yellowish green mate prevailed over his timidity and he alighted on a branch near her, while she rose up in the nest, redoubled her calls, and beat her uplifted wings into a haze. Without further delay, the male advanced to the nest's rim, to regurgitate food from his crop and pass it to his mate in several installments, while she continued to vibrate her expanded wings. I likewise found the nest above the Río Reventazón by seeing the male approach to feed his mate while she incubated her four eggs. In Colombia, Miller (*loc. cit.*) found three sets of 3 eggs and one set of 2. The dates of these sets, 24 May, 11 July, 19 December, and 4 January, suggest two breeding seasons in a year, as in the Rufous-collared Sparrow in the same locality.

Like related species, Lesser Goldfinches permit the nestlings' excreta to accumulate on the nest's rim, which may have a heavy white deposit by the time the young fly. The cleanness of the interior of the cup, however, indicates that the parents are not wholly neglectful of sanitation, but remove the droppings at least until the nestlings are strong enough to rise up and deposit them on the rim. What a contrast between the carelessness of goldfinches and the fastidiousness of most passerine birds, which not only remove all excreta from the nest but even carry off droppings that happen to fall upon surrounding foliage! Some, indeed, go so far as to dispose of the droppings of fledglings that have recently left the nest.

LARGE-FOOTED FINCH

Pezopetes capitalis

The Large-footed Finch is a stout, plain-colored sparrow about seven and a half inches in length. The sexes are alike in plumage, but the male is somewhat larger than the female. The forehead, forepart of the crown, lores, cheeks, and throat are black, which color extends, as two broad lines backward from the crown, into the dark gray of the hindhead, the back and sides of the neck, and the auricular region. The remainder of the body and the wing coverts are olive-green, lighter below; the remiges and rectrices are dusky with olive-green edges. The bill, the eyes, the legs, and the large, strong feet are black or nearly so.

This finch, the only species in its genus, is endemic in the highlands of Costa Rica and adjoining parts of Panama, where it ranges vertically from about 7,000 feet, or rarely a little lower, up to the bushy growth on the mountaintops at nearly 11,000 feet. In March of 1936, when I walked from El General to Cartago along the old packhorse trail that ran for a long way along the crest of the Cordillera de Talamanca (the route now followed in part by the Inter-American Highway), my attention was from time to time arrested by a loud rustling of the dead leaves which covered the ground at this dry season, and I paused long enough to glimpse these finches. They were especially abundant in the dense thickets of bamboos beneath open stands of tall oaks, 9,500 to 10,000 feet above sea level. Although they usually remained well hidden, I more than once saw them rummaging in the ground litter by raking it backward for several inches with both feet together. It was difficult then, as on later occasions when I watched these finches with more time to spare, to learn just how they managed this; but it reminded me of the similar foraging procedure of towhees and Fox Sparrows. The Large-footed Finch is the only member of its family that I have seen scratching with its feet in Costa Rica. Several species of *Atlapetes,* including the Chestnut-capped Brush-Finch and the Gray-striped Brush-Finch, use their bills to flick aside fallen leaves.

The Large-footed Finches that I saw along the trail over the Cordillera de Talamanca in March were mostly in pairs, as was to be expected at this season. I likewise found these finches in pairs high on the Volcán Irazú in late November; evidently they remain mated through the year and never associate in flocks. Here on Irazú they sometimes emerged from the brush-filled ravines to forage on the adjoining open pastures. They advanced over the ground by hopping with their feet together rather than by walking. At the lowest as at the highest altitude at which I have found them, these finches show a decided preference for dense stands of cane-like bamboos. At La Giralda, I met them occasionally as low as 6,700 feet in small patches of light woods with bamboo undergrowth, that had been permitted to grow in deep ravines amid the pastures.

In addition to the insects, worms, spiders, or whatever else they find in the ground litter, these finches eat berries, such as those of shrubby melastomes, which they rise a few yards above the ground to gather.

Voice

The song of the Large-footed Finch is an amazing performance, wholly different from that of any other sparrow that I have heard. One morning in early April, I listened with delight to one of these birds singing amid the mist-shrouded bamboos that formed the chief undergrowth of the high forest at about 7,500 feet. His song was a unique medley, consisting of a series of widely separated whistled notes, now high, now low in pitch, sometimes double or slurred, and often suggesting a thrush or an oriole (*Icterus*) with a slightly hoarse voice. From time to time, the songster unexpectedly interjected among his disjoined whistles a dry rattle, a loud chatter, or a trill that was almost soft and clear, arresting attention by the startling contrast. Black-faced Solitaires and Ruddy-capped Nightingale-Thrushes were at the same time singing in the surrounding woods, and the finch seemed to have taken musical hints from both of these gifted minstrels, although he did not succeed in reproducing their exquisitely modulated inflections.

On a later occasion, I counted the disjoined notes or phrases that this Large-footed Finch delivered in one minute, and found the number to vary from 14 to 20 when he was singing with greatest animation. He performed on a perch several yards above the ground, and after singing a while he flew down to forage among the fallen bamboo leaves, in the manner already described. For an hour, spells of singing alternated with spells of feeding. On some days, such singing would continue until past the middle of the morning. The songster was difficult to glimpse amid the close-set bamboos, and warily flew away the moment I came in sight of him.

Often I climbed up to the beautiful oak forest where this finch sang. He was at his best in April. In May he performed less freely. I spent most of 5 June in a blind in his territory, watching the nest of a Black-faced Warbler. The finch sang a good deal in both the morning and afternoon, but not so finely, I thought, as earlier in the season; there were fewer trills. By the end of June, he sang sparingly, and I heard only a few brief snatches in the course of several hours of the morning.

The Large-footed Finch has also a very different song. While traveling along the top of the Cordillera de Talamanca in March, I sometimes heard these birds give a performance which I then described as "a falling flutter of metallic notes." It reminded me of one of the songs of the Rusty-crowned Ground-Sparrow that I had heard in Guatemala not long before. In a small tract of woods in a ravine at La Giralda, on a morning in mid-June, I saw a Large-footed Finch fly down a steep slope, to vanish in a patch of dense, low herbage. Then, from amid this lush vegetation, I heard a loud, high-pitched, somewhat shrill *pee pee pee pee,* followed immediately by a much lower *chu chu chu,* then by some still lower chattering notes, the whole forming a stirring, not unmelodious song. I had repeatedly heard such a performance in the past and suspected that it belonged to *Pezopetes,* but this was my strongest—although still not unassailable—evidence for ascribing it to this bird. Probably it is a greeting song, given by both members of a pair when they meet after a brief separation, as in the case of the Yellow-thighed Finch.

Nesting

I searched assiduously for a nest in the bamboo thicket where I repeatedly heard the Large-footed Finch sing; but hunting for a nest amid such dense growth was like looking for a needle in a haystack, and finally I abandoned the quest as hopeless. The only record of the nesting of this finch that I have found is that given by Carriker (1910:895-896) in the following short account: "According to my observations on Irazú, the breeding begins from the 1st to the 10th of April. Four nests were found, one of the 10th, containing one egg partially incubated, and three on the 16th, two containing one egg and the third two eggs, all partially incubated. One egg seems to be the usual clutch, but one nest out of four having two eggs, while of the set of two, but one was marked, showing that the pigment was exhausted on the first egg laid. The nest is bulky, loosely built, and constructed of weed stalks and bamboo leaves, lined with soft blades of grass and placed in a thick bush or on a bamboo spray not far above the ground. The eggs vary from dull white, with a faint tinge of blue, to pale bluish, sparsely speckled and dotted over the whole surface with a few markings of lilac and heavier blotches of burnt umber or sooty, thicker at the larger end, and sometimes entirely wanting at the smaller. Average measurements: 29×19.4 mm."

YELLOW-THIGHED FINCH

Pselliophorus tibialis

The Yellow-thighed Finch is a slender, graceful bird about seven inches long. Except for the conspicuous lemon-yellow thighs, its plumage is everywhere dark: black on the pileum; sooty slate-color on the rest of the upper parts; dull slate below, with blackish throat. The bill is black; the eyes, legs, and feet are dark. The sexes are alike.

This easily recognized finch is confined to the highlands of Costa Rica and extreme western Panama. Its altitudinal range is from about 3,500 feet to timberline, and it is one of the most abundant birds in the wilder parts of Costa Rica throughout the wide altitudinal belt between 5,000 and 10,000 feet. Adults go in pairs throughout the year, sometimes in company with other small birds such as brush-finches and warblers, but never, in my experience, in flocks of their own kind. Active, versatile birds, they live not only in the heavy forests but likewise in bushy openings, and they frequently venture forth into shady pastures. They forage for insects and spiders as readily on the ground as in trees, and especially in wet, gloomy weather one may see them hunting through the pasture grass at a good distance from cover. If disturbed, they fly back to the woods or thickets where they feel most at ease. In trees they often ascend high to hunt through the foliage like wood warblers.

Yellow-thighed Finches vary their diet with berries, from which they carefully remove the skin, and they have interesting special methods of feeding. In humid subtropical forest I watched a Yellow-thighed Finch pluck a long, tubular, yellow corolla from the acanthaceous shrub *Jacobinia aurea,* squeeze the nether end in its bill, then drop the flower. The bird repeated this procedure with several corollas, evidently to press nectar from the base of each tube. Years earlier, in the Guatemalan mountains, I had watched Common Bush-Tanagers treat the crimson corollas of *Salvia nervata* in much the same fashion, obviously to procure the abundant nectar of this shrubby mint. When jays, orioles, Chisel-billed Caciques, and other icterids desire the nectar from a tubular flower, they generally pluck it and hold it beneath a foot; and with their sharp bills orioles may split the base of the flower to extract the sweet fluid through the slit. Finches and tanagers, however, seem not to use their feet for holding, but skillfully manipulate objects with the bill alone.

I have also watched Yellow-thighed Finches pluck the little white protein corpuscles (Müllerian bodies) from the brown hairy bases of the long petioles of *Cecropia* leaves. Small, sharp-billed birds like wood warblers, Bananaquits, and Red-faced Spinetails often gather these dainty tidbits, but the Yellow-thighed Finches are the largest birds that I have seen eat them. These protein corpuscles are, as is well known, the preferred food of the *Azteca* ants whose myriads inhabit the hollow stems and branches of the spindly *Cecropia* trees, which exhibit a number of anatomical peculiarities that make them favorable abodes for the ants. But in the highlands where the Yellow-thighed Finches dwell, these trees are usually devoid of ants, which are evidently deterred by the water that often fills the hollow internodes.

In its versatile foraging habits, its excursions into open fields and ascents to the treetops, *Pselliophorus* differs greatly from the species of *Atlapetes* that I know. The latter remain far more consistently in dense cover and on or near the ground, over which they hunt by flicking leaves aside with their bills.

Voice

In voice, too, the Yellow-thighed Finch contrasts with species of *Atlapetes* such as the Chestnut-capped Brush-Finch and the Striped Brush-Finch, for, although far from being a brilliant musician, its songs are decidedly more varied and cheerful. It has two songs, or, to be more exact, two types of songs, so different in character that familiarity with one would never lead you to attribute the other to the same bird. The first song, which is heard at all seasons of the year and all hours of the day, but more frequently as the breeding season approaches, is a long-continued, rapid flow of somewhat tinkling notes, all very much alike, with no definite phrasing. When first I heard this utterance, it reminded me of the songs of the Chestnut-capped and Gray-striped Brush-finches, although its tone was less squeaky. I finally assured myself that this tinkling song is delivered by both members of a pair, chiefly when they come together after a temporary separation. It is, in fact, a greeting song, an animated salutation, similar in its occasions to that of the Black-striped Sparrow, but more musical.

The songs of the second type are given chiefly at dawn in March, April, and May, and are rarely heard much after sunrise. Probably they are restricted to the males. These songs are shorter, structurally more complex, and quite different in tone from the first kind; they are repeated over and over, as the greeting song rarely is. These dawn songs vary greatly in phrasing from individual to individual. Typically they are short, rapid, breezy, little ditties, delivered in a peculiar high, dry tone. On April mornings at Montaña Azul, a Yellow-thighed Finch used to sing *Tee tiddie dee dee wink wink*. A neighbor of this finch had a shorter song, which fitted the words *Pity me sweet*. After he had repeated this many times over, he added another syllable and sang *Pity me sweet sweet*.

At La Giralda in 1963, I listened much to a Yellow-thighed Finch who lived in the woods beside a pasture. On the morning of 19 March, he began at 5:28 to perform within the edge of the woods.

Diddidcherchercherchup he sang very rapidly, in a thin, rather sharp voice, which increased in volume toward the end of each phrase. Each song lasted about one second, and from nine to 12 (usually 11) such songs were delivered in one minute. After a while, the songster came out of the woods into a hedgerow which led from it between two pastures, where he continued to chant as he moved from tree to tree, always away from his starting point. After he had travelled about 200 feet along the hedgerow, I heard him no more. It was then 5:46 a.m. In 18 minutes he had sung his lighthearted, breezy song 185 times, or at the rate of slightly more than 10 songs per minute.

Before sunrise one morning at the end of March, I found a Yellow-thighed Finch singing in the undergrowth of the forest, near its edge. Perching about 10 feet up, he tirelessly repeated a phrase that sounded like *t'chip-didichichi,* uttered very rapidly. His voice was softer and mellower than that of the other Yellow-thighed Finches I have heard. In his high-pitched, rapid, bizarre utterances, no less than in the restriction of his singing (other than the greeting song) largely to the interval before sunrise, the Yellow-thighed Finch reminds one of an American flycatcher rather than a finch. I am familiar with no other fringillid with a special song for the morning twilight.

Nesting

At Montaña Azul in 1938, the Yellow-thighed Finches began to nest in early March, but unfortunately the two nests that I found were completed before I noticed them. At La Giralda on 1 May 1963, I discovered a bird building a nest near the end of a long, descending, lower bough of a large cypress tree, at a point nine feet above the ground and well hidden by the foliage. This tree stood just above the head of a little wooded ravine largely overgrown with tall bamboos, amid extensive pastures. The finch directed my attention to her nest by carrying billfuls of dry bamboo leaves from the ravine up to her nest. I watched about five trips, on each of which the bird with material in her bill was followed by a mate who carried nothing. On the following morning, the builder was bringing fibrous material for the lining instead of leaves. The work was proceeding slowly, with only six billfuls being taken to the nest from 7:00 to 8:00 a.m., nine billfuls from 8:00 to 9:00, and none in the following quarter-hour. The nest was so well screened by the boughs of the cypress tree that observation was difficult; but as far as I could see, only one bird was building. Her material was still brought from the ravine, from which she reached the nest by a circuitous course that gave her the maximum protection from neighboring trees and led into the center of the nest tree, although to have approached the nest from the outside would have been more direct. Despite the building bird's evident concern for secrecy, she seemed to ignore my unconcealed presence on the grassy slope about 20 feet from the nest. Her mate sometimes escorted her back and forth, but he contributed nothing, as far as I could learn. Frequently I heard the tinkling greeting song (whether given by one or both partners, I could not tell), but never the very different song that is delivered chiefly at dawn.

The first of the two nests found at Montaña Azul was supported among densely matted elephant grass which leaned over the edge of a bank, beside a path that ran between the pasture and the adjoining forest. The structure was 3 feet above the edge of the bank and 7 feet above the pathway. Although well concealed below, it was rather exposed above. On 31 March, two four-day-old nestlings vanished from this nest, and on 18 April we found a replacement nest about 50 feet from the first. This nest amid the tall, coarse pasture grass was farther back from the bank and only 2½ feet above the ground. It already contained two eggs, which hatched a week later. These nests were bulky open cups, composed of straws and grass blades, with a thick, soft lining of fine, light-colored vegetable fibers.

Each of the three nests that I saw contained two eggs, on 15 March and 18 April 1938 and 7 May 1963. These eggs were white, or faintly tinged with blue, with a heavy crown of mingled brown and lilac speckles on the thicker end, and brown speckles thinly scattered over the remaining surface. The eggs of two sets measured 24.6 by 17.9 and 24.6 by 18.3; 25.4 by 18.3 and 24.6 by 18.3 mm.

Carriker (1910:897) found two nests "on the Volcan de Irazú at an altitude of about 9,000 feet, on April 14 and 16, 1902, containing one and two eggs respectively, and both with incubation begun. The nests were built entirely of dry bamboo leaves, lined with fine stems of grass, and placed on sprays of bamboo from ten to fifteen feet above the ground in deep thickly wooded ravines. Eggs pale bluish, thickly speckled, spotted, and blotched with lilac and chestnut-brown, more heavily about the larger end, forming a patch or wreath. Average measurements: 23.8 × 18.5 mm."

On her first nest in the elephant grass, the Yellow-thighed Finch always sat firmly on her eggs. Standing in the pathway below her nest, where she could not see me, I usually had to shake the nest a good deal to make her leave, so that I could see whether her eggs had hatched. If I pushed up a mirror, or my hand, I could nearly touch her before she hopped from the nest. On leaving it, she flew down into the bushes on the opposite side of the path, paused there a moment or two, then pushed through the thick foliage at the forest's edge and vanished.

On the afternoon of the day when her nestlings hatched, the mother was even bolder. She did not rise from her nest until I touched her tail with my uplifted hand. Then she turned to face me with one wing elevated in a defiant attitude. Next she dropped into the path almost at my feet and hopped across it toward the woods with her wings raised above her back, obviously trying to lure me away; but she did not "feign injury" in a convincing manner. From the shrubbery across the path, she flew back to the grass in which her nest was hidden. All this while she repeated a sharp note of distress almost incessantly.

A few days later, these nestlings vanished. In the replacement nest 50 feet away, which unfortunately we did not find until incubation was far advanced, the two eggs hatched on 24 April. By 3 May the nestlings were nearly feathered, and on 6 May, when 12 days old, they left their nest. Young Yellow-thighed Finches, when well feathered, rather closely resemble their parents, except that their thighs are dark gray like the abdomen.

CHESTNUT-CAPPED BRUSH-FINCH

Atlapetes brunneinucha

For a bird that lurks obscurely in the dark undergrowth of the forest, the Chestnut-capped Brush-Finch is exceptionally elegant. It is a stout finch, about seven and a quarter inches in length, and the sexes are alike. The black band that extends from the forehead around the eyes to the ear coverts is relieved by a white spot in the center of the forehead and one on each side. Above this band the pileum is chestnut, and the rest of the upper parts are plain olive-green with darker wings and tail. The chin, throat, and center of the breast and abdomen are white, with a black band across the chest. The sides are slaty-gray; the flanks olive-green; and the under tail coverts light olive-greenish. The short, stout bill is black; the eyes brown; the legs and feet dark brown.

The Chestnut-capped Brush-Finch is distributed through the mountains from central Mexico to Peru and Venezuela. In Mexico it has been found at altitudes ranging all the way from 1,200 to 11,500 feet (Miller *et al.,* 1957:346-347). In Guatemala, unaccountably, it seems to remain higher, from about 6,500 to 10,000 feet. In Costa Rica, I found this finch sparingly as low as 3,600 feet near Cañas Gordas on the Panamanian border, and at 4,700 feet on the northern slope of the Cordillera Central; but mostly it lives above 5,000 feet, and has been reported to range up to timberline (Carriker, 1910:899). In Venezuela its vertical distribution is similar, and it occurs from about 3,280 to 10,170 feet (1,000 to 3,100 m.) (Phelps and Phelps, Jr., 1963:420).

Chestnut-capped Brush-Finches forage over the ground amid dense bushy growth, either in the midst of the forest or in neighboring clearings choked with shrubby vegetation, especially in narrow valleys and ravines where little sunlight enters. In Costa Rica, they dwell beneath tall, epiphyte-laden subtropical forests, where they prefer areas with the thickest undergrowth; and I have found them also in lower secondary woods on cloud-bathed summits, where trunks and branches are green with mosses and larger epiphytes and the dense, tangled lower growth is hard for a man to penetrate. In Guatemala, the brush-finches inhabit Temperate Zone forests of oaks and other broadleafed trees mixed with pines, and likewise the almost pure stands of great cypress trees on some of the higher ridges. Everywhere they remain well hidden amid the bushy growths and are difficult to see.

Long ago, I passed a morning watching a hummingbird's nest, on a slope densely covered with young cypress trees near the summit of a high ridge in western Guatemala. While I sat quietly in my blind, a handsome bird emerged from the shrubbery behind me, alighted on the branches of a fallen dead tree directly in front of the blind, and turned from side to side as, with chestnut crest half raised and throat feathers expanded into a gleaming white puff, it scrutinized the strange object with one eye and then the other. Although in the preceding weeks I had enjoyed two or three fleeting glimpses of these strikingly attired birds, this was my first satisfactory view of a Chestnut-capped Brush-Finch. The attitude in which this bird scrutinized my blind was characteristic. Often, when surprised, a brush-finch will pause on a low perch and pivot nervously from side to side while it regards the intruder with suspicion. Even amid the dark undergrowth, the puffed-out snowy throat feathers make the bird conspicuous.

Chestnut-capped Brush-Finches never flock but live in pairs throughout the year. At the end of the breeding season, a pair may be accompanied by one or two offspring in juvenal or transitional plumage; but I have not seen more than two adults together. Mated birds are inseparable, and one rarely finds them more than a few yards apart, whether in flight or foraging amidst a thicket. Sometimes a pair of Chestnut-capped Brush-Finches keeps company with a pair of Yellow-throated Brush-Finches, birds almost as secretive as themselves. In the Costa Rican mountains, a pair of Chestnut-capped Brush-Finches may also associate with a pair of Highland Wood-Wrens and a pair of Black-cheeked Warblers.

Whenever I have succeeded in surprising a Chestnut-capped Brush-Finch as it foraged, it was hopping over the ground, flicking fallen leaves from side to side, or else pushing them forward, with vigorous movements of its bill. I could never come close enough to learn just what the bird found beneath the ground litter. Once I watched a brush-finch hovering on the outskirts of a swarm of army ants, doubtless catching the small insects that fled the hunting swarm. With it was a rare Barranca Finch.

Voice

The high-pitched, squeaky song of the Chestnut-capped Brush-Finch is a most inferior performance, hardly worthy of so large and handsome a finch. It resembles the song of the Yellow-throated Brush-Finch but is more forceful and continuous. On the Sierra de Tecpam in Guatemala, I did not become aware of the Chestnut-capped Brush-Finch's song until May, about the time the rains returned, although I had been studying birds in the same locality since the beginning of the year. Their singing increased as the weather grew wetter; it was most profuse in June, when the song of the majority of the birds was waning. I heard them occasionally, especially at daybreak, until early November. In the wetter Costa Rican mountains, brush-finches sing, as they nest, earlier than in the western highlands of Guatemala, where the dry season is long and severe. At Montaña Azul, the brush-finches began to repeat their melancholy song at dawn about the middle of February.

The call note is a weak, high-pitched *chink* or *pink*.

Nesting

On the Sierra de Tecpam in 1933, the majority of the small birds began to nest around the beginning of April, about the time when the last nocturnal frosts were seen at dawn on the open fields at 8,500 feet above sea level. The dry season was then at its height; but many trees had put forth fresh foliage during the preceding arid months, and they apparently supported an abundance of insect life. Those birds that foraged chiefly on the ground, including Ruddy-capped Nightingale-Thrushes, White-breasted Blue Mockingbirds, and Chestnut-capped Brush-Finches, waited longer, timing their breeding so that the nestlings would not hatch before the rains, which began in mid-May, had soaked the earth and favored the multiplication of the small creatures that live in or beneath the ground litter. The most advanced nest of the brush-finch that I found held young which left on 7 June; the eggs from which they hatched had evidently been laid around 10 May. In two other nests on the Sierra de Tecpam, eggs were laid in the second half of May; and the last of the four nests that I discovered there held well-incubated eggs on 20 June. This nest was situated in the cypress belt around 9,500 feet. On this cold height, conditions apparently became unfavorable for rearing the young as the wet season advanced, for in mid-July I found the two nestlings, who were beginning to be feathered, lying dead in their nest. The same disaster overtook several late broods of other species of birds.

In Costa Rica, Carriker (1910:899) recorded four nests with eggs between 16 and 19 April. At La Giralda, at an altitude of 7,200 feet, I found a nest in which eggs were laid in mid-April of 1963. Below Vara Blanca, at 4,900 feet, we discovered a nest with a single egg on 8 August 1938. These dates suggest a much longer breeding season than I observed on the Sierra de Tecpam, where the majority of the birds of all kinds had time to rear only a single brood. Near Soloma, on the northern side of the great rampart of the Sierra Cuchumatanes in the Department of Huehuetenango, Guatemala, Baepler (1962:152) found a nest with two well-incubated eggs on the surprisingly early date of 28 March. Here the dry season appears to be less severe than on the Sierra de Tecpam; "from November to May, rains are not infrequent and occur about once a week" (*op. cit.*, p. 141).

The six nests that I have seen were from 2 to 8 feet above the ground. Four of these nests, on the Sierra de Tecpam between 8,500 and 9,500 feet above sea level, were amid dense vegetation near woodland rather than in it. One nest was not far above the stream that flowed through a deep, narrow valley covered with bushes and scattered trees, where it was fairly well concealed in a low tangle of stems of a melastomaceous shrub. The second nest was in a thicket of blackberry bushes, also near a stream, in another deep valley. The lowest nest was 2 feet up in a low, dense shrub on a bushy hillside, near the edge of scrubby woods. The highest nest was 8 feet above the ground in a cypress sapling, one of a dense stand of these young trees, not far from a stream.

My two Costa Rican nests were in quite different situations. One was in the midst of the forest, 4½ feet up in a sapling, where it was screened by the small leaves of a slender climbing aroid. The second nest had been built near the end of a slender, drooping stem of a *Piper* shrub growing on the precipitous side of a narrow wooded ravine. It was about 5 feet vertically from the ground, but on such a slope a measurement of height means little. This nest was excellently concealed amid the large, clustered leaves of the *Piper*. It, too, was near the stream that flowed through the ravine, but in April when the nest was occupied this watercourse was dry. The narrow ravine, overgrown with shrubs and small trees, was situated amid extensive pastures with scattered trees. In this region, at the western end of the Barba massif, such a wooded ravine was often inhabited by one or a few pairs of several species of woodland birds.

The nest of the Chestnut-capped Brush-Finch is a bulky open cup. The Guatemalan nests had a foundation of broad dead leaves, or of decaying herbaceous stems and fibrous roots, or of leaves and weed stems and small twigs. In some nests there was a thick inner layer composed almost wholly of dry pine needles, with a few horsehairs in the bottom of the cup; one nest had a thicker lining of horsehairs; and another was well lined with black fibrous roots. The very bulky nest in the Costa Rican ravine had an untidy foundation consisting largely of long, coarse, dry herbaceous stems, some of which projected far beyond its walls. The thick middle layer was composed chiefly of long, slender, light-colored dead leaves of bamboo. The lining was of dark rootlets. This nest was 8 by 9 inches in over-all diameter by 5 in height. Inside it was 3 inches in diameter by 2¼ inches deep.

Four nests contained each two eggs or nestlings. One nest held a single egg which hatched a few days later. Another nest with a single egg was too remote to be revisited; possibly the set was incomplete. Carriker (1910:899) recorded three nests with single eggs (possibly representing incomplete sets) and one nest with two eggs. The pure white shells are slightly glossy. The measurements of eight eggs in Guatemala and Costa Rica average 25.7 by 18.9 mm. Those showing the four extremes measured 28.6 by 20.2, 25.4 by 20.6, and 23.4 by 17.5 mm.

At the nest in the deep ravine, the incubation period was 14 or 15 days, about the same as that of the Gray-striped Brush-Finch. At a nest in Guatemala, where the second egg was laid on 24 May, a parent continued to incubate, at least sporadically, until 12 June, after which I no longer found a bird present nor the eggs warm. I then opened these eggs, which the parent had continued to attend for 19 days, and found no trace of an embryo in either.

Newly hatched nestlings have pink skin shaded by long, dark gray down, rather abundant for a finch. When I removed two feathered nestlings for examination, their cries attracted the two parents, who crept over the ground and crawled through the bushes very close to me, quivering their wings slightly and incessantly repeating their sharp metallic *pink*. Knowing that the nestlings would jump from the nest if I replaced them while still excited, I held them until they became calm. When at last I laid them in the nest, they settled down contentedly; but when I returned to peep into the blackberry tangle and see if they were still present, a quarter of an hour later, the nest was empty. Doubtless the parents had called the fledglings away. At another nest, where the two young were not touched, the nestling period was 13 days.

When feathered, the nestling brush-finches are nearly everywhere almost uniform dark olive, with no trace of the chestnut crown, black face, and white throat which give their parents such a distinguished appearance. They have not yet acquired the more varied plumage, suggestive of the pattern of the adults but with duller colors, described for young birds by Ridgway (1901:465). The fledgling's upper mandible is nearly black, with yellow edges, and the lower mandible is mostly yellow. The iris is dark brown. This dull dress is soon exchanged for the elegant adult attire. On the Sierra de Tecpam on 23 July, I found two young birds still with their parents but already far advanced in the acquisition of the adult coloration. On both, the crown, and especially the hindneck, were noticeably tinged with chestnut; the cheeks were black; and some white feathers at the sides of the throat contrasted conspicuously with the olive around them. The black bar across the chest was already evident. Two months later, on 18 September, I found two young still bearing conspicuous traces of immaturity. After the middle of October, I could no longer distinguish the young of the year from the adults. The Chestnut-capped Brush-Finches that I saw on the Sierra de Tecpam were now so often in pairs that I suspected that the latest generation had not only acquired the full adult plumage but had also taken partners.

YELLOW-THROATED BRUSH-FINCH

Atlapetes gutturalis[1]

The Yellow-throated Brush-Finch is about six and a half inches in length and, as is usual in its genus, the sexes are alike in plumage. The top and sides of the head are uniform black, broken by a narrow white stripe that passes over the center of the crown to the hindneck. The rest of the upper parts are plain dark olive. The chin and throat are brilliant yellow or sometimes orange. The breast and abdomen are dull white in the middle, grayer on the sides. The bill is black; the eyes brown; the legs and toes dusky.

This attractive finch inhabits the highlands from extreme southern Mexico, in the state of Chiapas, to Colombia. At the northern extremity of its range, in Guatemala and adjacent parts of Mexico, it resides chiefly from about 5,500 to 10,000 feet above sea level. As happens with a number of highland birds, it stays at lower altitudes in southern than in northern Central America. In Costa Rica, it has been found from 2,000 to 8,000 feet and is most abundant between 3,000 and 6,000 feet. Here it is one of the few common birds on the thickly settled Central Plateau, where it lurks in dense hedgerows and thickets, bushy woodlands, weedy coffee plantations that are lightly pruned, and even shady gardens with abundant close-set shrubbery. In the wilder parts of Costa Rica, it inhabits the thick marginal undergrowth of heavy subtropical forest and adjoining clearings choked with weeds and bushes. In the Guatemalan highlands, I found Yellow-throated Brush-Finches in the undergrowth of open woodland composed of pines, oaks and other broadleafed trees and, on the higher mountaintops, amid dense stands of young cypress trees on cut-over slopes. Although by day they are usually found on or near the ground, in the evening they ascend into a tree to roost.

Like its congeners the Chestnut-capped Brush-Finch and the Gray-striped Brush-Finch, the Yellow-throated Brush-Finch remains paired throughout the year, as I have seen repeatedly in both Guatemala and Costa Rica. The mated couple are inseparable and forage close together, sometimes in company with a pair of Chestnut-capped Brush-Finches, but never, in my experience, near a second pair of their own kind. However, the young may remain with their parents after they have become difficult to distinguish, making family groups of three

[1] Paynter (1964) believes *A. gutturalis* to be conspecific with *A. albinucha,* and points out that the latter name has priority.

or four, which evidently break up long before the following breeding season.

I have found Yellow-throated Brush-Finches even more secretive than the Gray-striped and Chestnut-capped species. From what I have heard when approaching a pair of them, I am fairly certain that they forage on the ground, pushing the fallen leaves and litter aside with rustling sounds, doubtless with their bills in the manner of other brush-finches; yet I have never succeeded in watching these exceedingly wary birds while they engaged in this activity. They vary their diet with berries, of which I have seen them eat those of *Fuchsia arborescens*.

Brush-finches of the genus *Atlapetes* seem to depend on the striking color patterns of their heads and necks for mutual recognition more than on their voices. The Yellow-throated Brush-Finch is no better songster than its relatives that I have heard. As though ashamed of his squeaky voice and halting delivery, he does not, like his neighbor in the Guatemalan mountains, the Spotted Towhee, and many other ground-dwelling finches, rise to a high perch to carol, but he performs from an inconspicuous perch amid the bushes, so that he is almost as hard to see while he sings as at other times. Although, in the Guatemalan highlands in 1933, I began to hear the Yellow-throated Brush-Finch's song at the height of the dry season in late March, more than a month passed before I succeeded in tracing this distinctive utterance to its source. Of all the bird songs that I knew, this seemed to me the most expressive of an utterly broken heart. In a high-pitched voice, thin and slow, the finch seemed to lament:

O see me, O see,
I'm weary, pity me.

At the beginning of June, when two weeks of cold rains had silenced most of the songsters and the Yellow-throated Brush-Finch had little competition at daybreak, I heard this individual or a neighbor sing more forcefully and continuously. Nevertheless, his voice was so mournful, his tempo so slow, that it was hard to resist the impression that he greeted the chill, gray, misty dawn with the dirge that seemed appropriate to it. Until late in July, I continued to hear the brush-finches' plaintive songs.

Nesting

On a bushy hillside with scattered trees, at an altitude of about 8,500 feet on the Sierra de Tecpam in Guatemala, I found my first nest of the Yellow-throated Brush-Finch on 5 May, 1933. It was 2½ feet above the ground in a low tangle of vegetation at the base of a shrub. The bulky open cup, composed largely of pine needles with a few broad leaves and dry herbaceous stems in the outer wall, was sparingly lined with thinner pine needles and fine grass stems, the latter only in the bottom. The over-all diameter of the structure was 6 inches and its height was about 4½ inches. Internally, it measured 2½ inches in diameter and 2⅛ inches in depth.

On 25 June 1937, I discovered a nest in a sequestered corner of a shady garden on a coffee plantation near Cartago, Costa Rica, at an altitude of about 4,500 feet. The compact open cup, made chiefly of grasses, was 3½ feet up in a dense hedge of clipped bamboo.

In the following year, two nests were found in a narrow pasture surrounded by tall subtropical forest at Montaña Azul, at an altitude of about 5,300 feet. The first of these nests had been built among dense, low weeds, where it was so well hidden that the man cleaning the pasture chopped it down with his machete before he noticed it. It was a bulky open cup of straws, dry weed stems, and grass blades, with a lining of brown fibers and an additional lining of fine whitish fibers in the bottom. We propped the nest up as best we could in what we judged to be its original position, then replaced an egg that had rolled out and, as I learned later, had slightly cracked. The finch continued to incubate, hatched a nestling from her intact egg, and attended it until it fledged. The second nest was 10 inches above the ground among creeping cuicuyo grass (*Pennisetum clandestinum*), close beside a small stump, where it was shaded by sprouts from the stump and the frond of a fern. This bulky nest was composed of straws and a few leaf skeletons, and lined with rather coarse vegetable fibers.

The Guatemalan nest contained two fresh eggs on 5 May 1933. In the nest near Cartago, Costa Rica, three eggs were laid, on 26, 27, and 28 June 1937. The first nest found at Montaña Azul contained two eggs, one of which hatched on 6 April 1938; the second held three eggs, all of which hatched on 13 and 14 May 1938. These few records suggest that sets of two and three occur with about equal frequency. The eggs were pure white or had a faint tinge of blue. One of the ten which I have seen bore a few exceedingly fine brown specks; the others were immaculate. The measurements of nine eggs average 23.7 by 17.9 mm. Those showing the four extremes measured 24.6 by 17.9, 24.2 by 18.7, 22.6 by 18.3, and 24.2 by 16.7 mm.

At the nest amid the pasture grass, the parent sat until I came quite close, then jumped from her eggs and crept away through the grass, invisible to me, for several yards, before she took wing. This pair succeeded in rearing two of their three nestlings, who left the nest when 12 days old and well feathered. From the other nest in the same pasture, the single nestling left at the age of 12 days. The young brush-finches rather closely resembled their parents, but the stripe over the head was less definite and tinged with buff instead of being pure white, and the under plumage had conspicuous dark streaks, which the adults lack. These birds in juvenal plumage already had yellow throats, but paler than in the adults.

Like a number of other ground feeders, the Yellow-throated Brush-

Finches in the Guatemalan highlands nested later than the majority of the birds. As late as 19 July I noticed, at an altitude of about 9,500 feet, a fledgling who had apparently been out of the nest only a few days.

RUFOUS-COLLARED SPARROW

Zonotrichia capensis

This attractive finch, well known to Costa Ricans as the *Comemaiz* (maize-eater), is about five and a quarter inches in length. In both sexes, the top of the slightly crested head is marked by five parallel stripes that extend from the base of the bill to the occiput: a broad central stripe of gray; on each side of this, a black stripe; then a gray superciliary stripe. Below this is a narrow black stripe behind each eye, above the gray ear coverts. The chin and throat are whitish, set off by black patches on the sides of the chest that sometimes meet in the center. On the back and sides of the neck is a broad rufous collar which extends to the chest below the black. The upper parts, behind the rufous collar, are brownish, broadly streaked with black on the back. The brownish wings have two narrow whitish bars on the coverts. The breast and abdomen are white in the center, shading to grayish brown on the sides and flanks.

This sparrow, which, as pointed out by Chapman (1940), is of northern origin, has spread over the highlands of Middle America from the Mexican state of Chiapas to central Panama, and over the length and breadth of South America, from Colombia to Tierra del Fuego and from the Pacific Ocean to the Atlantic. It resides also in the mountains of Hispaniola in the Greater Antilles and on the small islands of Curaçao and Aruba in the southern Caribbean Sea. Even within the tropics, its altitudinal range extends from near sea level, as on Curaçao and the coasts of Peru and Brazil, up to 15,000 feet in the Andes. In Central America, it lives chiefly in the highlands, from 3,000 to 11,000 feet or more on the higher summits; but I found it abundant at Turrialba, a small town at only 2,000 feet on the Caribbean slope of Costa Rica.

To my great surprise, I awoke in Puerto Limón on the morning of 17 June 1935 with the familiar song of the Rufous-collared Sparrow in my ears. Looking around incredulously, I beheld one of these birds perching on the wheel that operated the hand-brakes of a freight car standing in the railroad yard. Two months later, I again heard Rufous-collared Sparrows singing in the railroad yard at Puerto Limón. I surmised that, somewhere along the higher reaches of the railroad line, which climbs to 5,000 feet, some of these sparrows entered a freight car to feed on spilled grain or some other food, were confined there when the doors were closed before the car moved off, and so were transported to the seaport no more than a hundred miles from their original home. This, at least, seems the most probable explanation of their presence so far from their usual habitat in Costa Rica. I have no recent information about Rufous-collared Sparrows at Puerto Limón. Aside from this, the lowest point at which the species has been recorded in Central America appears to be 1,600 feet, at Chamé in central Panama, where a single sparrow was seen on a sedge-covered hilltop (Rogers, 1945).

Through much of Latin America, the Rufous-collared Sparrow is a familiar town bird, by its abundance in the centers of human population reminding travelers from Europe or North America of the House or English Sparrow. On closer acquaintance, the visitors from the north are favorably impressed by its gentler, less aggressive ways and its far more melodious voice. Although in Central America the Rufous-collared Sparrow is established only in the highland towns, in South America it is abundant in such lowland cities as cool, cloudy Lima and warm, sunny Rio de Janeiro. In cities, the Rufous-collared Sparrow frequents parks, dooryards, patios—wherever it finds a little vegetation amid which to forage and shrubbery where it can take refuge from cats and boys with catapults, and build its nests. By no means confined to towns or even villages, it spreads over all the surrounding cultivated country, where it inhabits gardens, plantations, hedgerows, and pastures, and often swarms about the farm buildings. At suitable altitudes it lives in open country far from human habitations, but it strictly avoids closed forest and rarely ventures into the denser second-growth thickets.

This sparrow forages chiefly on or near the ground, over which it hops, gathering fallen seeds and capturing small invertebrates. According to Slud (1964:390), it "sometimes scratches industriously with both legs at once." Although it seeks the neighborhood of mankind, it is usually wary and avoids exposing itself. Its talent for concealing itself in the scantiest cover doubtless serves it in good stead in communities where laws for the protection of birds are lacking or loosely enforced, and to strike them down with pebbles propelled by rubber catapults is a too frequent pastime of small and often larger boys.

On the nearly level, cultivated plateau at the base of the Sierra de Tecpam at about 7,000 feet above sea level, Rufous-collared Sparrows were by far the most abundant birds. Indeed, they so swarmed in the hedgerows between the grain fields, and among the trees and bushes fringing the streams, that at times they appeared to be as numerous as all the other resident birds together. In the zone of oaks and pines halfway up the Sierra, around 8,500 feet, these sparrows were abundant about the buildings but far from common farther afield. In the clearings in the cypress forests near the summit, I did not find them even around the buildings, at 9,600 feet. But in the solitudes of the

Sierra Cuchumatanes, they dwelt among the scattered pine and alder trees even at 11,000 feet.

In localities where Rufous-collared Sparrows are only moderately abundant, it can be seen that they remain mated throughout the year; the pair that lived in the garden at Montaña Azul were inseparable even in November and December, when the breeding season was months away. But where very many of these sparrows are present, they congregate in such numbers in hedgerows and about farm buildings that pairs are not evident. Although gregarious, they do not travel in flocks, but scatter in different directions when disturbed.

Song

The first time I heard the pensive song of the Rufous-collared Sparrow, one sunny morning on the Sierra de Tecpam, I looked for a meadowlark, and was surprised to find the sparrow singing in a hawthorn tree. Others have noticed the resemblance of the Rufous-collared Sparrow's song to that of the Common Meadowlark. The song that reminded me of the meadowlark consisted of long double whistles that suggested gladness tinged with melancholy: *tseu-tseer, tseu-tseer.* Later I noticed other versions, such as *wheer whĕu, whĕu, wheer whĕu whĕu,* and *wheet wheet whēēēr,* several times repeated—all consist of long, clear whistles, sounding sweetly sad, that often bring to mind the White-throated Sparrow's song.

After a while, when I extended my travels and found the Rufous-collared Sparrow so abundant in so many places, it seemed to me that he sang *O, we're here, too, too* (the *we're* long drawn out, the *too, too* in a falling voice)—as though to remind us of his presence in all the higher towns and farmlands over many degrees of latitude and longitude. When I went to Ecuador, the first bird song that I heard as I awoke in Riobamba, after my first night in the Andes, was the Rufous-collared Sparrow's; and I was impressed by the resemblance of its plaintive whistles to those that I had recently heard in Costa Rica. When in full song, Rufous-collared Sparrows repeat their querulous notes all day long—and at times they break the stillness of the night with a few sweet, sad notes. They are more songful than most of the finches with which they live. Nearly always they sing from a low perch, such as a fence, a stalk in the garden, a shrub, or a lower branch of a tree.

In various parts of Central America, I have heard Rufous-collared Sparrows singing freely from early February until about the end of September, and in some years even later; but the amount of song is by no means uniform throughout this long interval. At Montaña Azul, they rarely sang in January of 1938 but came into full song by mid-February. Then song waned until in June there was little, but in July it increased again; in this month the sparrow was one of the most tuneful birds of the neighborhood. In August of the preceding year, Rufous-collared Sparrows sang all day, despite the wet, gloomy weather. At La Giralda in 1963, I noticed that song increased from a minimum in early June to exuberant singing, even on the most inclement days, before the month's end. These fluctuations in the volume of song are significant in relation to the breeding seasons.

Nesting Seasons

Although Rufous-collared Sparrows were abundant about most of the houses in which I dwelt while studying highland birds, I must confess that, absorbed in discovering the secrets of more colorful and retiring feathered creatures farther afield, I never gave these familiar neighbors the attention they deserved. However, I recorded such of their nests as came to my attention, and from these records there emerged the interesting conclusion—later confirmed by other observers—that the Rufous-collared Sparrow is one of the few Central American birds known to have two breeding seasons each year.

In Guatemala, between 7,000 and 11,000 feet above sea level, I found five nests, one of which contained eggs in May, one in June, and three in August. At Soloma, in the Department of Huehuetenango, Guatemala, Baepler (1962:152) likewise noticed two distinct nesting seasons, "the principal one in fall, when corn fields were the preferred habitat, and a lesser one in spring when the species frequented pastures and the shrubbery in town." In June he found two nests with young; spring nesting stopped by mid-July; then singing increased again and fall nesting started in mid-August. Between 19 August and the end of September, Baepler found four nests with eggs.

Below Vara Blanca in Costa Rica, on 2 August 1937, at the end of one of the dreariest spells of weather that I have known in Central America, we found three nests with eggs, some of which had been laid in the preceding month. No more nests were discovered until the following year, when we found one nest in which eggs were laid at the end of February and another where they were laid in early March. By April, when most other species of birds were just beginning to nest, Rufous-collared Sparrows were feeding fledglings on all sides. Although field work was continuous until my departure from Montaña Azul in mid-August, no more nests of this sparrow were found; but on 7 July 1940 a nest with eggs was discovered in this locality. At La Giralda in 1963, we found two nests in which eggs were laid in February, two in March, one in April, none in May, and one in which the eggs were laid in late June. Taking the Vara Blanca and La Giralda records together, in 12 nests on the Cordillera Central of Costa Rica, between 5,000 and 7,000 feet above sea level, eggs were laid as follows: February, 3; March, 3; April, 1; June, 1; July, 3; August, 1. No other bird of the region showed a similar distribution of laying dates.

Although the evidence that in Central America the Rufous-collared Sparrow has two breeding seasons each year is not massive, it is consistent for different localities and different years, and it is supported by observations on singing. On the Cordillera Central of Costa Rica, the first breeding season began about three months earlier than in western Guatemala, a difference evidently related to the shorter, less severe dry season in the former region—at Vara Blanca in 1938, there was hardly any dry season. In late July, when the sparrows at Vara Blanca were starting their second breeding season, practically all the other birds, except hummingbirds and the Slaty Flower-piercer, had finished, or were finishing, their one nesting for the year. Likewise in the Guatemalan highlands above 7,000 feet, few birds of other kinds nest in August when the Rufous-collared Sparrows resume breeding. As far as we now know, a double-peaked breeding season is rare among Central American birds. The Variable Seedeater and the Yellow-faced Grassquit, whose main breeding period is in the early part of the rainy season from May to August, may nest again on a reduced scale as the wet season passes into the dry in December and January, at least in the valley of El General (Skutch, 1954). In this same region, several pigeons, including the Ruddy Ground-Dove, Blue Ground-Dove, White-fronted Dove, and Rufous-naped Gray-chested Dove, nest toward the end of the dry season and at the beginning of the wet season, in February, March, and April, and then they nest again, on a reduced scale, in the rainy months of July, August, and September, with little or no breeding in May and June (Skutch, 1964b).

In Colombia, on the crest of the Western Andes at 3½ degrees north latitude and at an altitude of 6,500 feet, Rufous-collared Sparrows bred throughout the year, with pronounced peaks of nesting centering in mid-January and mid-June. Both periods of abundant nesting occurred in the latter part of a wet period and at the outset of the succeeding drier period (Miller, 1962). In this region, each adult undergoes a complete molt, lasting about two months, twice in the course of a year. In the intervening periods of four months, they may breed. Females never molt while attending nests; if they start to nest while molting, the molt is arrested. Males breed chiefly, but not exclusively, between molts. At this low latitude, Rufous-collared Sparrows begin to breed at a very early age; one female started when only five months old (Miller, 1961). The relation between the breeding seasons of Rufous-collared Sparrows in Central America and their molts is not known. Whereas the intervals between peaks of nesting activity in the Western Andes of Colombia are respectively five and seven months, in Central America these intervals are less equal; but even during the shorter one, May or June to August in western Guatemala and March to July on the Cordillera Central of Costa Rica, there should be time for a complete molt.

Even beyond the tropics, as in the La Plata region of Argentina, the Rufous-collared Sparrow appears to have two distinct nesting periods each year. The statement of Hudson (1920, 1:56) that "two broods are reared in the season, the first in October, the second in February or March," suggests two separate breeding seasons rather than one that is long continued. The question merits further study.

Nest and Eggs

The 18 nests of the Rufous-collared Sparrow that I have seen in Central America were in varied sites. Five were in shrubs or stump sprouts, at heights up to 4 feet above the ground. Thorny growths with close-set branches, such as hawthorns browsed by cattle and unpruned rosebushes, were preferred. Two nests were at ground level, one of them among low herbage in a marshy meadow. On the stormy Cordillera Central of Costa Rica, by far the preferred site was in or beside a bank, either a roadside cutting or the side of a shallow ditch. Sometimes these nests were set in niches in moss-covered banks, where they were completely sheltered from rain; somtimes they were roofed by the overhanging roots and soil at the top of the bank; and sometimes they were hidden amid matted grass or other vegetation that draped over the bank and offered a measure of protection from the frequent rains of this region. One nest was on a beam in a cowshed, 8 feet above the floor. A similar diversity of nest sites is reported from other parts of this sparrow's wide range.

Except the nest on the beam in the cowshed, which was a loose, untidy accumulation of straws, the nests of the Rufous-collared Sparrow that I have seen were neat, well-made cups. These nests varied in composition, according to what each locality afforded. In most, the outer wall was made of coarse or fine grasses, often with an admixture of slender weed stems, thin dead vines, fibrous rootlets, and the like. Sometimes a little green moss was included in the wall: a nest in the midst of an alpine meadow on the high Sierra Cuchumatanes had a thick outer layer of green moss, a middle layer of fibrous roots and grass stems, and a liberal lining of hairs from the horses that roamed half wild over the high plateau. Where available, horsehairs are generally chosen for the lining, and often there are also downy feathers, usually from domestic chickens. One nest, set snugly in a niche in a mossy bank, was lined with white horsehairs and tawny, hair-like spore stalks of the moss *Funaria,* with the empty capsules still attached and sticking up above the rim at the front. The lining is sometimes of fine vegetable fibers, and if nothing more tenuous is available, the finest grasses are used.

A nest in a marshy meadow was roofed over by the living and dead grass blades which had evidently been pushed up to form a space for its reception. Only through a circular opening in the grasses at one side could the sparrow enter or leave her nest, which appeared to be an oven-shaped construction, although the bird had built only the usual open cup.

In Guatemala, three nests contained 2 eggs or nestlings and two nests contained 3 eggs or nestlings. In Costa Rica, I found nine sets of 2 eggs and two sets of 3 eggs. Beyond the tropics in Argentina, however, the *Chingolo,* as our sparrow is there called, lays 4 or 5 eggs, thus exhibiting the trend, common in many other families and species of birds, for clutch size to increase with latitude. In Central America, the Rufous-collared Sparrow's eggs are light blue, sometimes greenish blue, blotched and speckled all over with shades of brown, the markings usually most concentrated on the thick end, where sometimes they obscure the blue ground color. Twelve eggs, measured at the nest in Guatemala and Costa Rica, averaged 20.8 by 15.6 mm. Those showing the four extremes measured 21.8 by 15.1, 21.0 by 16.3, and 19.8 by 15.1 mm.

Although Rufous-collared Sparrows seek the vicinity of man because they find life easier there, they do not trust him; they hesitate long to approach their nests in his presence, even when he watches from a good distance away. When one examines their nest, the parents protest with a sharp, staccato note, repeated as long as the intruder remains there. In Guatemala, two nestlings raised in the same nest left when 10 days old. In Costa Rica, a solitary nestling remained in the nest for 12 days.

LITERATURE CITED

ARMSTRONG, E. A.
1963. A study of bird song. Oxford University Press, London.

BAEPLER, D. H.
1962. The avifauna of the Soloma region in Huehuetenango, Guatemala. Condor, **64**: 140-153.

BELCHER, C., AND G. D. SMOOKER
1937. Birds of the colony of Trinidad and Tobogo. Ibis, ser. 14, **1**: 225-249; 504-550.

BENT, A. C.
1953. Life histories of North American wood warblers. Bull. U. S. Nat'l Mus., **203**: 1-734.

BLAKE, E. R.
1956. A collection of Panamanian nests and eggs. Condor, **58**: 386-388.

CARRIKER, M. A., JR.
1910. An annotated list of the birds of Costa Rica, including Cocos Island. Ann. Carnegie Mus., **6**: 314-915.

CHAPMAN, F. M.
1917. The distribution of bird-life in Colombia: a contribution to a biological survey of South America. Bull. Amer. Mus. Nat. Hist., **26**: 1-729.
1940. The post-glacial history of *Zonotrichia capensis*. Bull. Amer. Mus. Nat. Hist., **77**: 381-438.

CHERRIE, G. K.
1916. A contribution to the ornithology of the Orinoco region. Sci. Bull. Mus. Brooklyn Inst. Arts Sci., **2**: 133a-374.

DAVIS, J., G. F. FISLER, AND B. S. DAVIS
1963. The breeding biology of the Western Flycatcher. Condor, **65**: 337-382.

DAVIS, T. A. W.
1949. Display of White-throated Manakins *Corapipo gutturalis*. Ibis, **91**: 146-147.

EISENMANN, E.
1962. Notes on some Neotropical vireos in Panama. Condor, **64**: 505-508.

FRIEDMANN, H., L. GRISCOM, AND R. T. MOORE
1950. Distributional check-list of the birds of Mexico. Part I. Pacific Coast Avif. (Cooper Orn. Soc.), no. 29, 202 p.

GILLIARD, E. T.
1959. Notes on the courtship behavior of the Blue-backed Manakin (*Chiroxiphia pareola*). Amer. Mus. Novitates, no. 1942, 19 p.

GRISCOM, L.
1932. The distribution of bird-life in Guatemala. Bull. Amer. Mus. Nat. Hist., **64**: 1-439.

HAMILTON, T. H.
1962. Species relationships and adaptations for sympatry in the avian genus *Vireo*. Condor, **64**: 40-68.

HUDSON, W. H.
1920. Birds of La Plata. 2 vol. J. M. Dent, London.

JOHNSON, N. K.
1963. Biosystematics of sibling species of flycatchers in the *Empidonax ham-*

mondii — oberholseri — wrightii complex. Univ. California Publ. Zool., **66**: 79-238.

LAND, H. C.
1962. A collection of birds from the Sierra de las Minas, Guatemala. Wilson Bull., **74**: 267-283.

MACQUEEN, P. M.
1950. Territory and song in the Least Flycatcher. Wilson Bull., **62**: 194-205.

MILLER, A. H.
1961. Molt cycles in equatorial Andean sparrows. Condor, **63**: 143-161.
1962. Bimodal occurrence of breeding in an equatorial sparrow. Proc. Nat'l Acad. Sci., **48**: 396-400.
1963. Seasonal activity and ecology of an American equatorial cloud forest. Univ. California. Publ. Zool., **66**: 1-78.

MILLER, A. H., H. FRIEDMANN, L. GRISCOM, AND R. T. MOORE
1957. Distributional check-list of the birds of Mexico. Part II. Pacific Coast Avif. (Cooper Orn. Soc.), no. 33, 435 p.

MITCHELL, M. A.
1957. Observations on birds of southeastern Brazil. Univ. Toronto Press.

MOORE, R. T.
1944. Nesting of the Brown-capped Leptopogon in Mexico. Condor, **46**: 6-8.
1947. Habits of male hummingbirds near their nests. Wilson Bull., **59**: 21-25.

MOYNIHAN, M.
1962. Display patterns of tropical American "nine-primaried" songbirds. I. *Chlorospingus*. Auk, **79**: 310-344.

MUMFORD, R. E.
1964. The breeding biology of the Acadian Flycatcher. Misc. Publ. Univ. Mich. Mus. Zool., no. 125, 50 p.

PAYNTER, R. A., JR.
1964. The type locality of *Atlapetes albinucha*. Auk, **81**: 223-224.

PHELPS, W. H., AND W. H. PHELPS, JR.
1958. Lista de las aves de Venezuela con su distribución. Tomo II. Parte I. No Passeriformes. Bol. Soc. Venezolana Ciencias Nat., **19**: 1-317.
1963. Lista de las aves de Venezuela con su distribución. Tomo I. Parte II. Passeriformes. Segunda edición. Bol. Soc. Venezolana Ciencias Nat., **24**: 1-479.

RIDGWAY, R.
1901. The birds of North and Middle America. Bull. U. S. Nat'l Mus., **50** (1): 1-715.
1902. The birds of North and Middle America. Bull. U. S. Nat'l Mus., **50** (2): 1-834.
1907. The birds of North and Middle America. Bull. U. S. Nat'l Mus., **50** (4): 1-973.
1914. The birds of North and Middle America. Bull. U. S. Nat'l Mus., **50** (6): 1-882.

ROGERS, G.
1945. Costa Rican Sparrow near Canal Zone. Auk, **62**: 641.

ROWLEY, J. S.
1962. Nesting of the birds of Morelos, Mexico. Condor, **64**: 253-272.

SCHÄFER, E.
1954. Sobre la biología de *Colibri coruscans*. Bol. Soc. Venezolana Ciencias Nat., **15**: 153-162.

Sick, H.
1959. Die Balz der Schmuckvögel (*Pipridae*). Journ. f. Ornith., **100**: 269-302.

Skutch, A. F.
1944a. The life history of the Prong-billed Barbet. Auk, **61**: 61-88.
1944b. Life history of the Blue-throated Toucanet. Wilson Bull., **56**: 133-151.
1946. The hummingbirds' brook. Sci. Monthly, **63**: 447-457.
1949. Life history of the Yellow-thighed Manakin. Auk, **66**: 1-24.
1950a. The nesting seasons of Central American birds in relation to climate and food supply. Ibis, **92**: 185-222.
1950b. Life history of the White-breasted Blue Mockingbird. Condor, **52**: 220-227.
1952. Scarlet Passion-flower. Nature Mag., **45**: 523-525; 550.
1953. How the male bird discovers the nestlings. Ibis, **95**: 1-37; 505-542.
1954. Life histories of Central American birds. I. Pacific Coast Avif. (Cooper Orn. Soc.), no. 31, 448 p.
1958a. Life history of the Violet-headed Hummingbird. Wilson Bull., **70**: 5-19.
1958b. Roosting and nesting of araçari toucans. Condor, **60**: 201-219.
1960. Life histories of Central American birds. II. Pacific Coast Avif. (Cooper Orn. Soc.), no. 34, 593 p.
1961. Life history of the White-crested Coquette Hummingbird. Wilson Bull., **73**: 5-10.
1962a. Life histories of honeycreepers. Condor, **64**: 92-116.
1962b. The constancy of incubation. Wilson Bull., **74**: 115-152.
1964a. Life history of the Scaly-breasted Hummingbird. Condor, **66**: 186-198.
1964b. Life histories of Central American pigeons. Wilson Bull., **76**: 211-247.

Slud, P.
1957. The song and dance of the Long-tailed Manakin, *Chiroxiphia linearis*. Auk, **74**: 333-339.
1960. The birds of finca "La Selva," Costa Rica: a tropical wet forest locality. Bull. Amer. Mus. Nat. Hist., **121**: 55-148.
1964. The birds of Costa Rica: distribution and ecology. Bull. Amer. Mus. Nat. Hist., **128**: 1-430.

Snow, D. W.
1956. The dance of the manakins. Animal Kingdom, **59**: 86-91.
1963. The display of the Blue-backed Manakin, *Chiroxiphia pareola,* in Tobago, W. I. Zoologica, **48**: 167-176.

Todd, W. E. C., and M. A. Carriker, Jr.
1922. The birds of the Santa Marta region of Colombia: a study in altitudinal distribution. Ann. Carnegie Mus., **14**: 1-611.

Wagner, H. O.
1945. Notes on the life history of the Mexican Violet-ear. Wilson Bull., **57**: 165-187.
1955. Zur Biologie des Bergclarinos (*Myiadestes obscurus* Lafr.). Bonner Zool. Beiträge, **3-4**: 200-206.

Worth, C. B.
1938. Nesting of the Pepper-shrike. Auk, **55**: 539.
1939. Nesting of some Panamanian birds. Auk, **56**: 306-310.

INDEX

COMMON NAMES OF BIRDS

INDEX

SCIENTIFIC NAMES OF BIRDS